A History of the Solar System

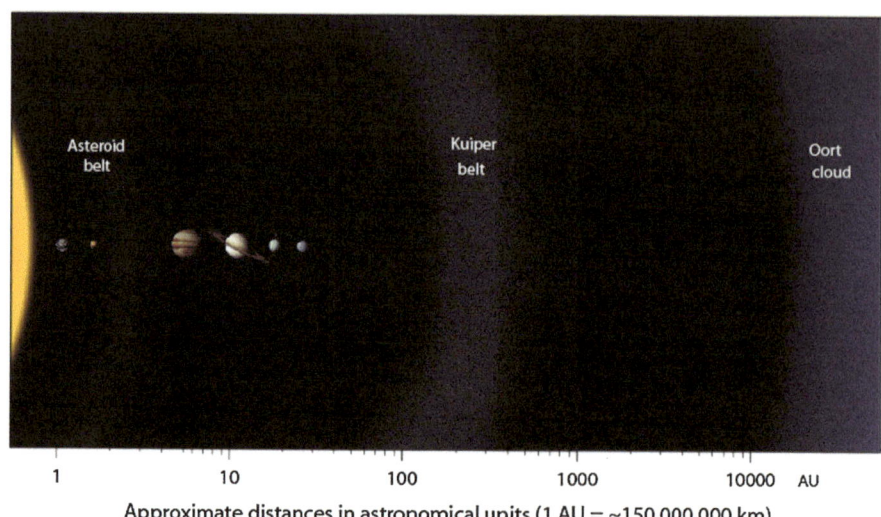

Approximate distances in astronomical units (1 AU = ~150,000,000 km)

The Oort Cloud, a hypothetical spherical reservoir at 10^3–10^5 AU, contains 10^{11} to perhaps 10^{12} comets; the disk-like Kuiper Belt, at 30–1000 AU, contains 10^8–10^9 comets; the asteroid belt contains 10^9–10^{12} asteroids

Claudio Vita-Finzi

A History of the Solar System

 Springer

Claudio Vita-Finzi
Department of Earth Sciences
Natural History Museum
London
UK

ISBN 978-3-319-33848-4   ISBN 978-3-319-33850-7   (eBook)
DOI 10.1007/978-3-319-33850-7

Library of Congress Control Number: 2016941089

Cover artwork © Don Dixon/cosmographica.com

Printed on acid-free paper

This Springer imprint is published by Springer Nature
The registered company is Springer International Publishing AG Switzerland

*Damn the Solar System! bad light—planets
too distant—pestered with comets—feeble
contrivance; —could make a better with
great ease*

*Lord Francis Jeffrey (1773–1850) according
to the Rev. Sydney Smith*

For Alexandre

Preface

As we move with ever greater confidence between the planets, their moons, a few comets and asteroids, and some grains of dust, and prepare to enter interstellar space almost 20 billion km from Earth after a journey of 36 years at 61,000 km/hr, it seems a good moment to consider the history of the only planetary system we are currently capable of exploring in any detail. But the discovery of over two thousand planets which are orbiting stars other than our own Sun will undoubtedly spur humanity before long to find ways of visiting those alien worlds in one way or another.

This short book outlines a story which spans 4.5 billion years and which is the fruit of a few millennia of naked eye observation, four centuries of squinting through telescopes, and sixty years informed by orbiting satellites and manned and unmanned probes and landers, profound laboratory studies, and imaginative hypotheses.

My principal aim is to link events dating back billions of years which we can glimpse among the stars with our everyday concerns on Earth and to demonstrate that the solar system continues to evolve and diversify. Although the chapters are broadly in chronological order, I have tried to get away from the 'one era after another' scheme by devoting successive chapters to a brief history of ideas about the solar system; the raw materials of which the solar system is constructed; their assembly into solid, gaseous, and icy bodies; the evolution of the solar system's key player, the Sun; the major changes undergone by the planets and moons after they had formed; the emergence of life; and some of the current changes that help us understand the solar system's past. Some of the material is difficult but so is the subject matter, and the drift will usually be clear from the context. Above all, I hope to have conveyed the excitement and wonder that comes from looking up—at the sky and in the library.

Every one of these themes draws on advances in geochemistry, biology, and computing as much as to targeted space missions and ground-based observation, and to the work of individuals, teams, and space agencies, in particular the ever generous NASA, debts I try to acknowledge in the references and captions.

I am grateful to Paul Henderson and his successors in London's Natural History Museum for hospitality, to Mark Biddiss, Ken Blyth, Louis Butler, Ian Crawford, Dominic Fortes, Kenneth Phillips, Michael Russell, Sara Russell, Fred Taylor, Leo Vita-Finzi, and Michael Woolfson for their searching but kindly comments on parts of the text; to Simon Tapper for help with the figures; to Don Dixon for the cover image; and to Petra van Steenbergen and Hermine Vloemans at Springer for support.

Ferrara, March 2016   Claudio Vita-Finzi

Note: Myr is used throughout for million (10^6) years and Gyr for billion (giga, 10^9) years

Contents

Chapter 1
Introduction

Abstract Thanks to the discovery since 1995 of multiple planets orbiting Sun-like stars we know that, as intuited by Giordano Bruno in 1584, our solar system is not unique. The nebular hypothesis for its origin, first clearly stated by Pierre-Simon de Laplace in 1796, has proved durable, while our understanding of its evolution, including the part played by contributions from other parts of the Milky Way galaxy, has been enriched by the geochemical analysis and dating of material from the Moon, Mars, meteorites and other solar system bodies as well as the Earth.

We now know that there are countless solar systems in the universe. The notion is not novel but one that long ran counter to religious and scientific dogma. The Dominican friar Bruno [5] suspected that there were many suns around which many earths revolved as did the planets around our Sun ('innumerabili mondi simili a questo'). In 1600 he was burnt at the stake for this and other deviations from Church doctrine.

Newton [26] also considered the possibility of 'other star systems': in the General Scholium at the close of his *Principia* he argued that 'if the fixed stars were the centres' of systems like ours they would be subject to the dominion of an intelligent and powerful being who had spaced them immense distances apart lest they crashed into one another under the force of gravity. Laplace [20] likewise spoke in his *Mécanique céleste* of 'the solar system and analogous systems scattered throughout the immensity of the heavens'.

The first recorded use of the term 'solar system' is thought [30] to date from 1704, but the sun-centred or heliocentric model in which it is rooted comes and goes, at least since Aristarchus of Samos (3rd century BC), and it does not finally mature until the publication of Copernicus' [14] *On the Revolutions of the Celestial Spheres* in which the Sun rules over the planetary family. Copernicus (like Galileo) implicitly equated his planetary system with the entire universe (Fig. 1.1), just as the Milky Way galaxy (Fig. 1.2) was subsequently considered to encompass the entire universe until 1917 when it was shown to be just one of 100,000,000,000

© Springer International Publishing Switzerland 2016
C. Vita-Finzi, *A History of the Solar System*, DOI 10.1007/978-3-319-33850-7_1

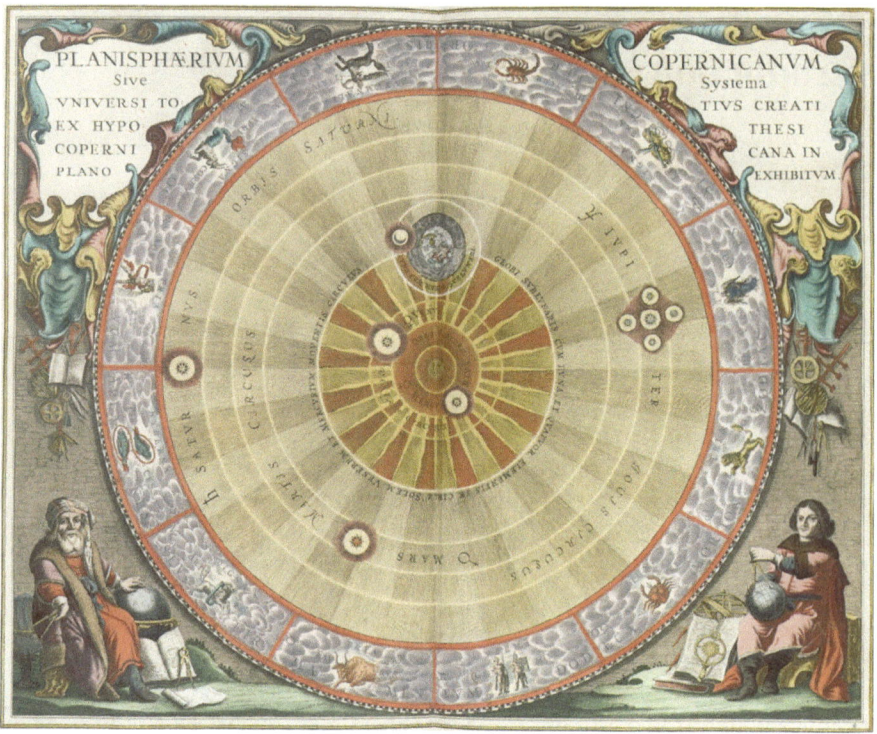

Fig. 1.1 Copernicus' Sun-centered system. Note the assumption that the entire universe is embraced by the planisphere. From Andreas Cellarius' *Planisphaerium Universi Totius Creati ex Hypothesi Copernicana in Plano*, 1660, From http://www.staff.science.uu.nl

galaxies. Our current concern with parallel universes shows that astronomers have absorbed that lesson. In England Thomas Digges not only popularised the Copernican model but added a vast panoply of stars beyond the original notion of a dome of fixed stars (Fig. 1.3).

The *Encyclopedia Britannica* of 1972 (i.e. a year after the launch of the first space station, Salyut 1) noted that over half the stars in the Milky Way were binaries or systems of higher multiplicity and that stars with companions $17\times$ the mass of Jupiter were known, so that it was reasonable to suppose that many stars were accompanied by bodies of planetary dimensions. But, despite a few such bold conjectures about the possibility of other solar systems, the evidence for a multiplicity of stars each with its planetary retinue is recent and still incomplete.

The key was of course the discovery of extrasolar planets. The first persuasive evidence of planets outside our solar system came in 1992 when two or more planet-sized bodies were found to be orbiting a pulsar (PSR B1257+12: [31]). In 1995 a large planet was found orbiting 51 Pegasi [23] which, like our Sun, is a main-sequence star. And by February 2016 over 2000 extrasolar planets were known, some forming part of 509 multiple planetary systems.

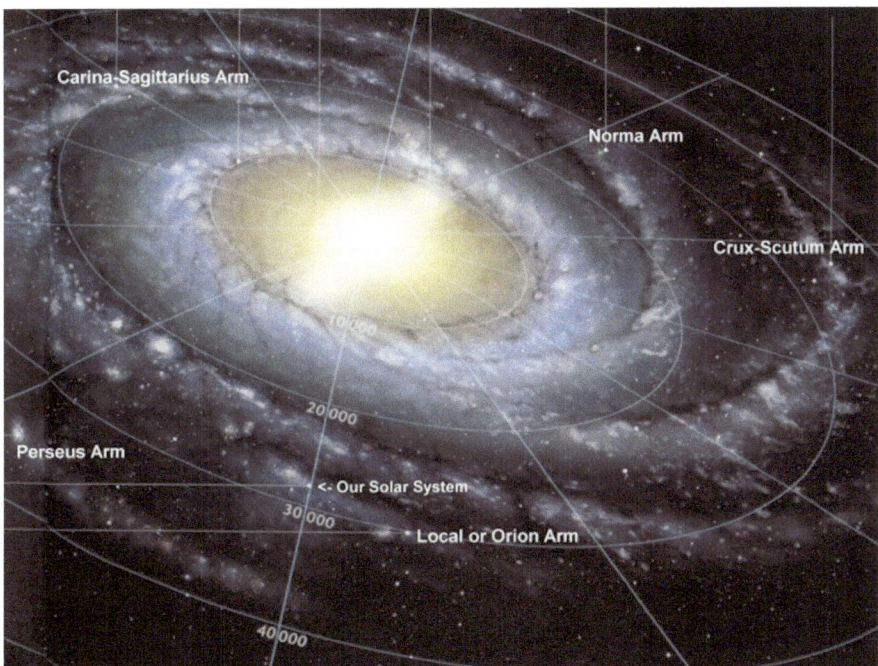

Fig. 1.2 Location of our solar system in the Orion Arm of the barred spiral Milky Way galaxy, estimated to contain 100–400 billion stars. The concentric scales are in light years. From http://www.universetoday.com

Exoplanets are identified mainly by the transit method (Fig. 1.4), where the passage of a planet temporarily reduces the brightness of a star, and by spectroscopy, which detects variations in the radial velocity of the star relative to the Earth resulting from shifts in the combined centre of mass of the star and its planet. The former method was used by the Kepler space observatory, launched in 2009, which monitored the brightness of 145,000 main sequence stars. By 7 January 2015 it had detected some 440 stellar systems and this led to an estimate of 11 billion Earth-sized planets orbiting Sun-like stars in the Milky Way.

The question remained whether the groupings amount to solar systems, not purely a matter of semantics but because the interplay between planetary bodies (as later chapters show) bears on such matters as the origins of life and the evolution of atmospheres. Four years of observation by the Kepler satellite revealed extrasolar system KOI-351, consisting of 7 transiting planets with orbital periods that range between 7 and 330 days and with the two innermost planets similar in size to the Earth [8] (Fig. 1.5). The familiarity does not stop there: the outer orbits are occupied by gas giants, and the planets interact dynamically in a manner akin to the phenomenon of orbital resonance named after Laplace and displayed by the moons of Jupiter, which occurs where two or more orbiting bodies have a mutual gravitational effect because the ratio between the period of their orbits is close to a small whole number, such as the values 1:2:4 identified by Laplace for Io, Ganymede and Europa.

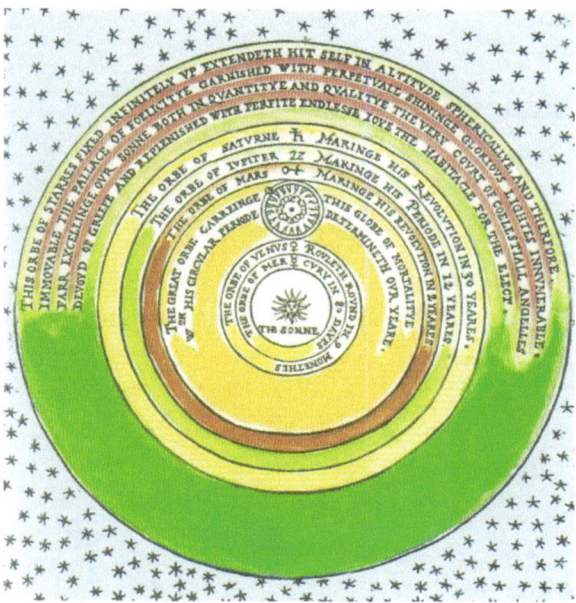

Fig. 1.3 Thomas Digges (c1546–1595) went beyond Copernicus by postulating an infinite universe of stars. Image in public domain

As the hunt for exoplanets was at least initially driven mainly by the search for life elsewhere in the cosmos the next step was to identify those planets that orbit their parent stars within the habitable zone. How this should be defined was of course problematic, but the presence of liquid water featured in most schemes on the grounds that it is an essential compound for the existence of life as we know it [25]. That is an unnecessarily restrictive target. In any case the presence of life is by no means an essential component of solar systems as broadly defined nor, indeed, something restricted to them.

As early as 1783 William Herschel wrote a paper on the 'Motion of the Solar System in Space' and soon afterwards he concluded that the Sun lay near the 'bifurcation' of the Milky Way, impressively close to current assessments of its passage through the galaxy's spiral arm to its location in the Orion Arm (Fig. 1.2). And as late as the 1970s it was still reasonable to define the solar system on the basis of its 'gravitational attraction' as 'extending to the orbit of the outermost known planet, Pluto, 40 astronomical units from the sun' (AU, the mean distance between Earth and Sun or 150 million km), but prudent to add that if, however, it is considered to extend to the aphelia of comets with nearly parabolic orbits, 'its extent is… approximately 100,000 astronomical units' (Encyclopedia Britannica 1972).

This is broadly how the solar system is still normally defined today [16] although authorities differ over whether the Oort Cloud should be excluded, as its members do not orbit the Sun except individually and accidentally. However, as shown in Chap. 6,

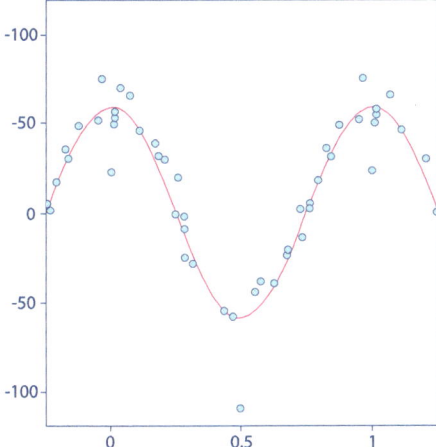

Fig. 1.4 The transit method for detecting extrasolar planets: periodic variations in radial velocity (in m/s: y axis) of the solar-type star 51 Peg (Gliese 882) indicating a companion with a mass of at least 0.5 that of Jupiter (M_j) orbiting at 0.05 astronomical units (AU). Based on Mayor and Queloz [23]; error bars removed for clarity. x axis: orbital phase (period 4.23 days), solid line denotes orbital motion of 51 Peg computed from orbital parameters

the gravitational influence of a large body in the outer reaches of the solar system [3], which would extend the newtonian limits of our solar system to 12,000 AU, well within the limits of 50,000–200,000 AU estimated for the outer Oort cloud [25].

Even so the need is increasingly felt for a definition which is in tune with the needs of space exploration or with the analysis of space climate. One such is the heliosphere, which, until recently largely of academic interest, has grown in significance with the problems for telecommunications created by space weather and for what it says about solar activity. It is the region of space dominated by the

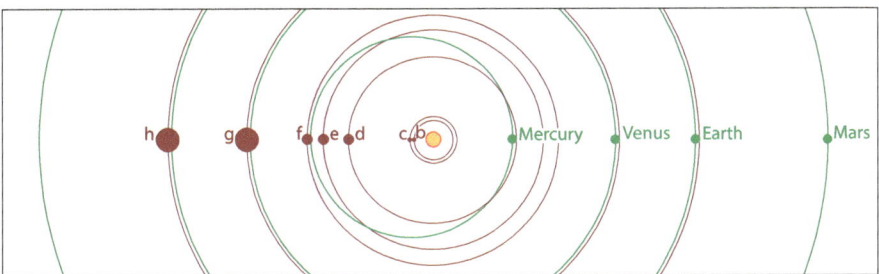

Fig. 1.5 KOI-351 was described as a solar system similar to ours but more compact as all its members are within 1 AU. It was identified by the Kepler Space Telescope and was the first system to be found with small, probably rocky, planets in the interior (b-f) and gas giants (h and g) in the exterior (Cabrera et al. [8]). Three of the planets have orbital periods similar to those of Mercury, Venus and Earth, but they vary by as much as 25.7 h. Credit DLR and Astrobiology 29.11.2013

Fig. 1.6 System of vortices in Descartes' Principia Philosophiae, Part 3 [15]. According to this scheme the Universe contains many systems of bodies revolving around central stars of which our solar system is one such whirl. The Sun in surrounded by ethereal material and when it rotates other planets are forced to orbit around it. S = Sun, R H N Q = ellipse

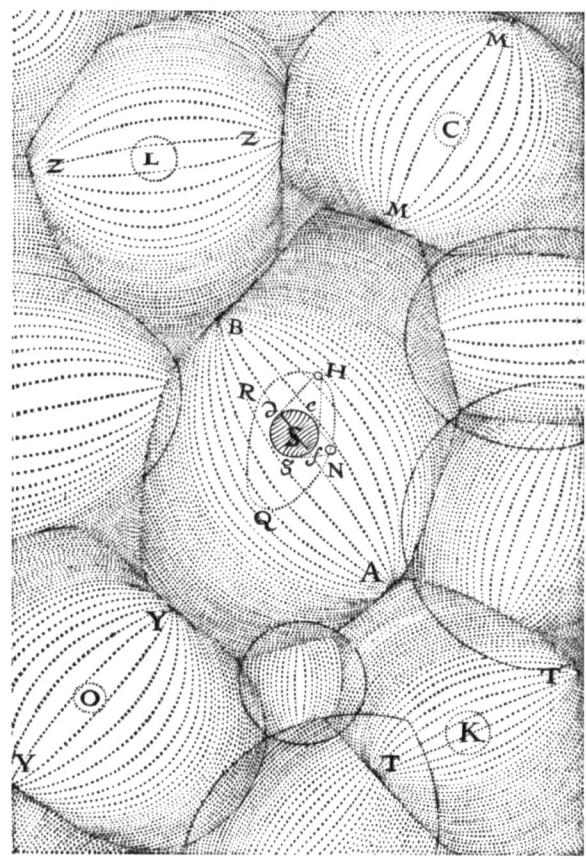

flow of ionised particles emitted by the Sun's corona—the solar wind—and which extends through the Kuiper belt to over 100 AU (Frontispiece). Another, related, definition of the edge of the solar system is the limit of the interplanetary magnetic field (IMF), which is carried by the solar wind [27]. Indeed, the heliosphere is sometimes defined as a magnetic field inflated by the solar wind.

In the cosmology of Descartes [15] the universe was filled with vortices composed of luminous, transparent and opaque elements; in his solar system the vortex had developed a series of stratified bands circling around the Sun at different speeds and each home to a planet (Fig. 1.6). A related suggestion made by Swedenborg [28] is that the proto-sun developed a dense crust which under the action of centrifugal force thinned and broke up into pieces—the eventual planets and smaller bodies (Fig. 1.7). Whatever its defects—and it is sometimes derided by association with the author's poetic and philosophical baggage—the proposal prefigured what we now call the nebular hypothesis, according to which a molecular cloud collapsed to form the Sun as well as a protoplanetary disc which gave rise to the bodies making up the solar system.

Versions of the scheme were later formulated by Count Buffon (George-Louis Leclerc) , Immanuel Kant and Laplace. Buffon [7] suggested the planets formed

Fig. 1.7 Planetary evolution according to Swedenborg [28]. The figure shows the outer zodiacal belt, its rupture, and the planets and their satellites migrating towards their orbits

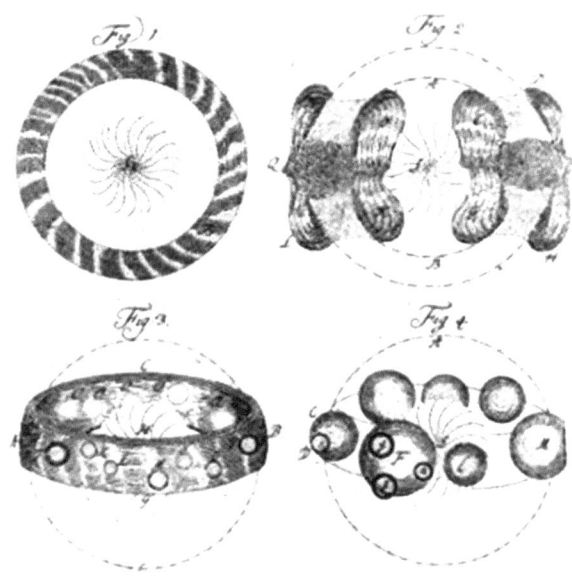

about 75,000 years ago from matter (*un torrent de matière*, as Laplace put it) torn from the Sun by a large comet. Kant [19] invoked gravity as the device by which particles in space came to orbit in the same plane, with the denser ones eventually forming the planets nearest to the Sun and the less dense forming the large, outer planets endowed with numerous satellites by virtue of their large masses [6].

Laplace [20] also relied on gravity for his model, in which rotation of a hot gaseous cloud produced a disk which contracted and therefore spun faster as it cooled to yield successive rings at its margin. These condensed to planets and their satellites, a notion consistent with what was known at the time about solar system bodies (7 planets and 14 satellites) and their orbits, though not with the concentration of the system's angular momentum in the planets (98 %) rather than in the Sun, which incorporates 99.8 % of the mass of the solar system. When Napoleon asked Laplace why God was missing from his *Mécanique céleste* he replied 'Sir, I had no need of that hypothesis'. The tale reflects Laplace's rejection of Newton's appeal to divine horology combined with his thoroughgoing determinism. But by 1773 (Fig. 1.8) there were already those who focused on the geometry of the solar system without much concern for its evolution.

Laplace's nebular hypothesis, as it came to be called, has proved remarkably resilient in the face of great advances in astronomy and physics and in our understanding of the composition of the early solar system that are owed to computational advances and sampling missions to planets, comets and asteroids as well as to the information derived from meteorites. It dominated the issue for a century, thanks in large measure to the authority of Herschel [18], who concluded that stars we can see lie in a thin but extensive plane above which he discerned 'a canopy of discrete nebulous masses, such as those from the condensation of which he supposed the whole stellar universe to have been formed'.

Fig. 1.8 Title page of Benjamin Martin's calculations (1773) for determining the dimensions of the solar system using solar parallax based on two transits of Venus. 'not only the most rare but also the most curious Phaenomenon of the Heavens.'

The nebular model was briefly challenged when improved telescopes seemed to show that nebulae were merely aggregations of stars, but spectroscopic observation soon showed that at least one nebula yielded a single spectral line rather than the multiplicity to be expected from several stars. By the same token spectroscopy showed that, with a few exceptions, nebulae, the Sun and the Earth were made up from the same family of elements [6]. The philosopher Comte [12] had declared that we would never be able to study the chemistry or temperature of the stars. 'Never' would seem as dangerous an adverb in astronomy as in politics.

The essence of the Laplacian scheme—rotation, flattening, differentiation— holds good in most of the many models that have since been proposed for the origin of the solar system. As Aitken [1] noted in 1906, 'The general consensus of opinion for more than a century has been that our Sun and its system developed into its present form from an earlier nebular state.' Aitken went on to summarise the work of Chamberlin and Moulton [10] outlining in 1900 a substitute based upon the assumption of an original spiral nebula. Spiral or discoid these authors still invoked a parent nebula. Much more recently, a classification of theories of planetary formation [16] which lists 39 different proposals, merely distinguishes between those in which planets formed from unaltered interstellar material as opposed to stellar material raised to high temperatures, and between models in which the Sun and planets formed at the same time rather than separately. All of them set off from a nebular disk. Similarly, the opening sentence of a review of research on the early Earth published in 2014 simply speaks of 'the processes that assembled dispersed dust and gas in the solar nebula' [9].

In surprising contrast, the 'prenatal' history of the solar system has been much refined on the basis of the composition and age of its constituents and observation of present-day nebulae and protostars. If the short version seems little changed from that of Laplace—collapsing molecular clouds 'swiftly evolve to form young stars surrounded by disks from which planets originate' [4]—we can now estimate the time taken for our stellar nursery to form. The secret is to measure the period during which radioactive isotopes have decayed to their present solar system abundances while cut off from the supply of radioactive material freshly synthesised within stars. One potential cosmochronometer, ^{182}Hf (hafnium182), yields ~30 million years (Myr) as the time required for the Sun's gestation [22], a modest value when set against the billions required for the rest of the system to be crafted.

Water assumes special prominence in this part of the narrative as a versatile clue to early non-biological chemistry. As discussed in Chap. 2, theoretical studies and chemical data suggest that the water in the most primitive objects in our solar system, namely minerals within meteorites (as envoys from asteroids) and comets, is older than the Sun (a child of the nebula) and points to a source in the parent molecular cloud. In other words, if our solar system is typical there are interstellar ices (and water for life) available for any number of other infant planetary systems [11].

The actual birth of the solar system has been pinned down by the radiometric dating of meteoritic inclusions to 4567.3 ± 0.16 Myr ago [13]. Most accounts speak of the ensuing *evolution* of the solar system, just as they did for the molecular clouds. Biologists and astronomers dispute their claims to priority in the adoption of the term, whether or not it is encumbered with the assumption of progress. Some claim that the nebular model accustomed geoscientists to think in terms of cumulative change; others give the credit for this to biology. It is a sterile dispute except insofar as 'evolution' might also imply increased complexity and therefore evidence of something more than differences in age.

Just how these claims can change is illustrated by a letter to *The Times* of London dated 12 April 1871 and signed Astronomicus which complained that 'Mr. Darwin's theory requires us to believe that animal life existed on this globe at a period when, according to a theory much more plausible than his, the earth and all the planets with the sun constituted but one diffused nebula'. Astronomers, he observed, had data at their disposal including changes in the configuration of several nebulae recorded in historical times, whereas the variations that Darwin points to, especially in man, are 'either zero or of an extremely nebulous character' (darwin-online.org.uk).

The balance in data content between astronomy and biology in the study of the early Solar System was restored once geologists had grappled with the constitution and age of the Earth independently of the demands of Darwinian evolution and developed a free-standing chronology, although it has failed to shed its historical attachment to the vagueness of Greek names in preference to numerical ages), witness current enthusiasm for the Anthropocene to define a slice of Earth history dominated by human activity [17].

Indeed, Earth history played an influential part in calibrating solar system prehistory, a consequence both of the key personalities and of technical progress. Today astrogeology, the field defined and energised by Eugene Shoemaker at the US Geological Survey, is officially concerned with planets rather than stars whereas astrobiology has from the outset spread its net more widely. But solar system astronomy remains heavily reliant on geoscience through meteorites and for geological evidence to supplement the meagre 400 years of telescopic history. The birth of the solar system is confidently proclaimed on the basis of a few microscopic samples that fell to Earth whereas its development is laboriously and tentatively pieced together from the deformation and disruption of an array of planets, moons and ices.

The ensuing chapters try to convey current thinking on the key events and processes. Many of the journal references cited here can be accessed at least in part via the internet; full-length texts which explore the subject thoroughly but from very different viewpoints include Alfvén and Arrhenius [2], Taylor [29], Lewis [21] and Woolfson [32]. New findings and novel interpretations are reported almost daily, but there remains a troubling traditionalism in a substantial part of the literature. To quote Lewis [21] writing two decades ago, 'We, in the late 20th century, still live under the shadow of the clockwork, mechanistic world view first formulated in the 17th century... We must internally turn our educations upside down to accommodate a universe that is quantum mechanical and relativistic, within which our "normal" world is only a special case'.

References

1. Aitken RG (1906) The nebular hypothesis. Pub Astron Soc Pacific 18: 111-122
2. Alfvén H, Arrhenius G (1976) Evolution of the solar system. NASA, Washington DC
3. Batygin K, Brown ME (2016) Evidence for a distant giant planet in the solar system. Astron J 151, 2
4. Bizzarro M (2014) Probing the solar system's prenatal history. Science 345: 620-653.
5. Bruno G (1584) Dell'infinito universo et mondi (On the Infinite Universe and Worlds). Venice
6. Brush S G (1996) Nebulous Earth. Cambridge Univ Press, Cambridge
7. Buffon G-L Leclerc (1749) Histoire naturale, générale et particulière. Impr Royale, Paris
8. Cabrera J et al (2014)The planetary system to KIC 11442793: a compact analogue to the solar system. Astrophys J 781, 18
9. Carlson O et al (2014) How did early Earth become our modern world? Annu Rev Earth Planet Sci 42:151-178
10. Chamberlin TC, Moulton FR (1900) Certain recent attempts to test the nebular hypothesis. Science 12:201-208
11. Cleeves LI et al. (2014) The ancient heritage of water ice in the solar system. Science 345:1590-1593
12. Comte A (1835) Cours de philosophie positive. Bachelier, Paris
13. Connelly JN and 5 others (2012) The absolute chronology and thermal processing of solids in the solar protoplanetary disk. Science 338, 651-655
14. Copernicus N (1543) De revolutionibus orbium coelestium. Petreius, Nuremberg
15. Descartes R (1644) Principia philosophiae. Elzevirius, Amsterdam
16. Encrenaz T, Bebring J-P, Blanc M (1990) The solar system (2nd ed) Springer, Berlin
17. Hamilton C, Bonneuil C, Gemenne F (eds) (2015) The Anthropocene and the global environmental crisis: rethinking modernity in a new epoch. Routledge, Abingdon

18. Herschel W (1811) Astronomical observations relating to the construction of the heavens. Phil Trans 101:269-345
19. Kant I (1755) Allgemeine Naturgeschichte und Theorie des Himmels (Eng trans 1981) Petersen, Königsberg
20. Laplace PS de (1798) Traité de mécanique céleste. Duprat, Paris
21. Lewis JS (1997) Physics and chemistry of the solar system (rev ed). Academic, San Diego
22. Lugaro M et al (2014) Stellar origin of the Hf-182 cosmochronometer and the presolar history of solar system matter. Science 345: 650-653
23. Mayor M, Queloz D (1995) A Jupiter-mass companion to a solar-type star. Nature 378: 355-359
24. Morbidelli A (2008) Comets and their reservoirs: current dynamics and primordial evolution. In Jewitt D et al (eds) Trans-Neptunian objects and comets. Springer,New York, 79-164
25. Mottl M et al (2007) Water and astrobiology. Chem Erde 67: 253-282
26. Newton I (1713) Philosophiae naturalis principia mathematica (2nd ed). Joseph Streater, London
27. Parker EN (2007) Conversations on electric and magnetic fields in the cosmos. Princeton Univ Press, Princeton
28. Swedenborg E (1734) Opera philosophica et mineralia. Hekel, Leipzig
29. Taylor SR (2005) Solar system evolution. A new perspective (2nd ed). Cambridge Univ Press, Cambridge
30. Webster Merriam (2015) Online dictionary accessed March 2014
31. Wolszczan A, Frail DA (1992) A planetary system around the millisecond pulsar PSR1257+12. Nature 355:145-147
32. Woolfson M (2015) The formation of the solar system: theories old and new (2nd ed). Imperial Coll Press, London

Chapter 2
Raw Materials

Abstract A presolar (molecular) cloud supplied dust and gas to the solar nebula ancestral to our solar system. The dust originated in different varieties of star including low mass stars at the end of their evolution and exploding supernovae. They include silicates and graphite. The gas is predominantly molecular hydrogen (H_2). Calcium-aluminium-rich inclusions (CAIs) and chondrules are found in meteorites which yield some of the oldest ages for the solar system. Polycyclic aromatic hydrocarbons (PAHs) are widespread; at low temperatures they may be transformed into more complex organic molecules by UV radiation.

The composition of the giant molecular cloud that was ancestral to our solar nebula is inferred from the residual nebula itself, from analogy with 'stellar nurseries' in other parts of the Milky Way and other galaxies, and from computer models. As with research on human origins when founded on genetic data, these procedures yield probabilities and not certainties; and, of course, the search is two-way: finding ancestors requires a sound grasp of the genotype (or its equivalent) of the descendants, in this case our particular ancestral planetary nursery, and of environmental change during the period at issue.

In the earliest phase of star formation, dense and cold cores of molecular clouds are supported against gravitational collapse by magnetic and turbulent as well as thermal effects. Magnetic fields within the cloud are enhanced by shearing caused by turbulence (Fig. 2.1). Theories that hinge on turbulence and on magnetic fields broadly agree in indicating that molecular clouds collapse on the way to forming stars on a time scale of ~10 Myr [26]. That molecular clouds form and fade in less than ~10 Myr is also suggested by evidence that the carbon monoxide (CO) in our galaxy is found mainly in its spiral arms, as this is the time required for cloud material to pass through the arms [35].

The collapse into a nebula is thought to take place once a core has attracted a critical mass of dust and gas from the parent cloud. Examples can be seen as the red clumps in the image of the Orion Molecular Cloud Complex (Fig. 2.1). The most prominent, to the lower left of centre, is the Orion Nebula, also known as M42 according to the 1774 catalogue of nebulae and star clusters by Charles

© Springer International Publishing Switzerland 2016

C. Vita-Finzi, *A History of the Solar System*, DOI 10.1007/978-3-319-33850-7_2

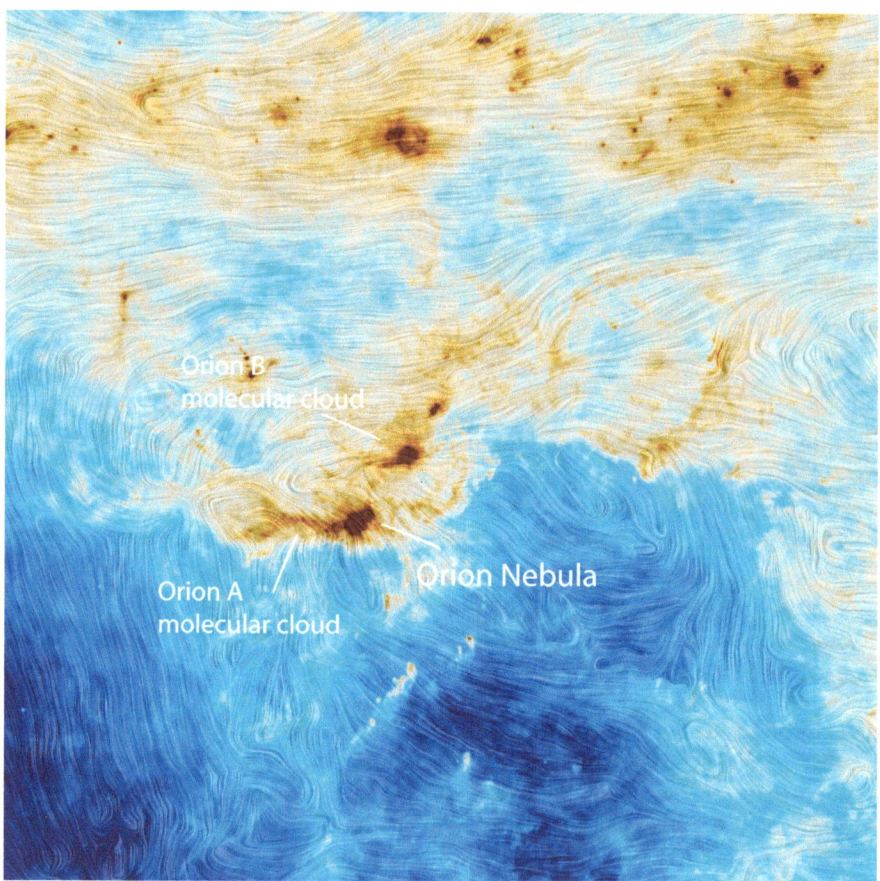

Fig. 2.1 Location of the Orion Nebula and of two molecular clouds. Image based on data from ESA's Planck satellite. Magnetic field orientation is shown by the texture, as in the upper part of the picture where the field is arranged roughly parallel to the Galactic plane. *Yellow* and *red areas* reflect denser and mostly hotter clouds containing larger amounts of dust and gas. Image roughly 40° across, Courtesy of ESA and the Planck Collaboration

Messier; at 1340 light years it is the nearest large star-forming region. The nebulae in turn generate protostellar discs each surrounding a protostar. Hubble has revealed 42 such protoplanetary discs or proplyds in the Orion area (Fig. 3.1). (The term protoplanetary nebula is confusingly also used for the penultimate stage in the evolution of an intermediate-mass star.)

Gas

Molecular clouds are dense and large enough to allow molecules—usually H_2—to form. They have temperatures in the region of 10 K (that is, a mere 10 degrees above absolute zero) and diameters of 20–200 parsecs (a parsec or pc corresponds

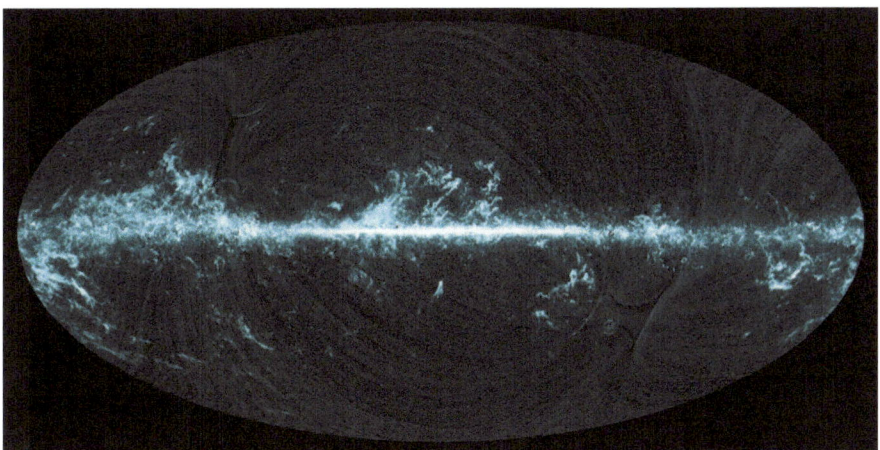

Fig. 2.2 Column density of molecular hydrogen in entire sky based on carbon monoxide probed by Planck's High Frequency Instrument (HFI). Courtesy of ESA/Planck Collaboration

to ~3.3 light years) and they measure up to 1 million solar masses. The discovery of molecular clouds in our galaxy has greatly progressed [23] to the point where chemical and physical composition, temperature and active processes may be observed in several examples. We are talking about features many light years distant, but the range of specimens—some 6000 in the Milky Way—is such that a composite history can be constructed, to the benefit of our parochial concerns, by relying on the principle (sometimes called ergodic) by which past states in one system are inferred and ordered from the present condition of several other analogous systems.

As cold H_2 does not emit much radiation we use other molecules for inspecting the clouds. CO has been found useful in this connexion as it emits several emission lines within the range of the frequency detectors on the Planck satellite (Fig. 2.2). Planck was launched by the European Space Agency in 2007 primarily to detect variations in the Cosmic Microwave Background radiation (CMB) in order to identify, among other things, changes in the wavelength of the CMB as a measure of expansion of the Universe. The CMB has been described as a snapshot of the oldest light in our Universe when it was just 380,000 years old. Molecular clouds are in the way of the CMB; their imaging was a byproduct of their analytical removal ('foreground subtraction'), not unlike the identification of individuals removed from images in successive Soviet encyclopedias.

A molecular cloud also shows up at infrared (IR) wavelengths because its molecular hydrogen is well mixed with about 1 % of dust consisting of silicates and graphites. Like interplanetary dust particles (IDPs) in general, those found in molecular clouds were formerly ignored, dismissed as a nuisance to observers, or appreciated for their collective delineation of the Horsehead Nebula (Fig. 2.3), but IDPs are now ascribed an important role in the evolution of galaxies, stars and planetary systems and in the synthesis of organic molecules.

By analogy with the Orion and other nebulae and their ancestral molecular clouds the Sun, like other stars, inherited its chemical composition from a nebula which was

Fig. 2.3 The Horsehead Nebula, part of the Orion Molecular Cloud complex, a dark cloud of gas and dust silhouetted against the bright nebula IC 434. *Credit and Copyright* Jean-Charles Cuillandre and Hawaiian Starlight (Canada-France-Hawaii Telescope in Hawaii) and NASA

itself derived from a molecular cloud. Hydrogen, helium-3 and helium-4, deuterium and lithium were produced by the nucleosynthesis associated with the Big Bang, whereas the metals (a term used by astronomers for elements heavier than helium and hydrogen) were created by stellar nucleosynthesis in stars which, on completing their stellar evolution, returned their material to the interstellar medium. Since the Sun formed, some of the helium and metals have settled under gravity from its photosphere (or visible surface); the protostellar Sun's composition, to judge from that of primordial (C1 chondrite) meteorites, had a higher metal content than today's 1.49 % [22].

Dust

IDPs may thus include grains whose origin antedates our solar system. These presolar grains generally consist of silicates and graphites emitted into interstellar space by stars within which they been generated by nucleosynthesis, that is to say the production of novel chemical elements by nuclear fusion, a process which sometimes imprinted the grains with a distinctive isotopic signature.

The isotopic analyses that reveal their history are made on grains as small as 1 μm, that is to say a thousandth of a millimetre, but they reveal a wide range of distinctive sources. They include Asymptotic Giant Branch (AGB) type stars, which are at a late stage of their evolution and emit substantial winds from which grains condense, and supernovae, which are explosions of a white dwarf in a binary star system or of a massive star which has exhausted its nuclear fuel [4]. In this way laboratory studies have added an important dimension to astrophysical observation and greatly extended the range of solar system history.

The silicates, which include crystals of two minerals commonly encountered on Earth, enstatite ($MgSiO_3$) and forsterite (Mg_2SiO_4), are viewed as a fundamental building block of solar systems. They have in fact been detected around both young and old stars by the Infrared Space Observatory (ISO) [8] as well as in chondritic meteorites and comets. Silicate dust absorbs UV and visible light that is emitted by stars in the vicinity, and emits it as IR radiation. The two indices are linked, as the extremely low temperatures that prevail in the cloud may favour the precipitation of CO on the dust grains.

As regards the graphite component, analysis of IDPs coupled with chemical modelling and astrophysical considerations show it was contributed by outflow from AGB stars of about 1.1–5 solar masses ($M_\odot$) [5]. In other wornebula and then trapped in clathratesds we now have an idea of the dimensions as well as the evolutionary stage of the source stars, an integral part of attempts to trace 'the timing and tempo of the transformation of the disk of gas and dust to the solids that formed the planets' in our solar system and by extension in other solar systems [11] and the extent to which stellar material is recycled from earlier generations.

The IR emission that betrays the presence of dust also reflects any heating by ultraviolet (UV) radiation and cosmic rays, and hence the level of resistance by the cloud to gravity. Gravitational potential energy was invoked in the 19th century, notably by Lord Kelvin and Hermann von Helmholtz, to power the Sun (leading to a maximum age for it of 31 Myr). In the present context the concept enters into the Jeans length (λJ), a measure of the relative strength of the gravitational force and the resisting gas pressure.

Many molecules besides H_2, carbon and silicates are found in nebulae. They range from simple molecules such as ammonia (NH_3) and CO to complex organic molecules. A spectral line survey of Orion nebular clouds by the Far Infrared (HIFI) instrument on ESA's Herschel satellite (launched in 2009) revealed (Fig. 7.1) water and sulphur dioxide (SO_2), and several organic compounds, among them formaldehyde (CH_2O), methanol (CH_3OH), dimethyl ether (CH_3OCH_3) and hydrogen cyanide (HCN). Polycyclic aromatic hydrocarbons (PAHs) have also been detected. The organic molecules have been described as 'potential life-enabling organic molecules'; the processes that might have fulfilled this promise to the ultimate benefit of terrestrial life are discussed in Chap. 7.

Ices

Ices are thought to make over half of the material which condensed in the solar nebula at about 4 AU from the Sun, notably the ices of water (Fig. 2.4), carbon dioxide, ammonia and methane [12]. As we have seen, dust and gas have been

Fig. 2.4 Phase diagram of
water to illustrate how minor
changes in temperature or
pressure may produce major
changes in behaviour, most
familiarly at the triple point
in the lower centre but more
subtly with the 18 known
ices. Phase transitions have to
be taken into account when
considering such matters as
the role of convection on the
larger icy moons. Courtesy of
various sources

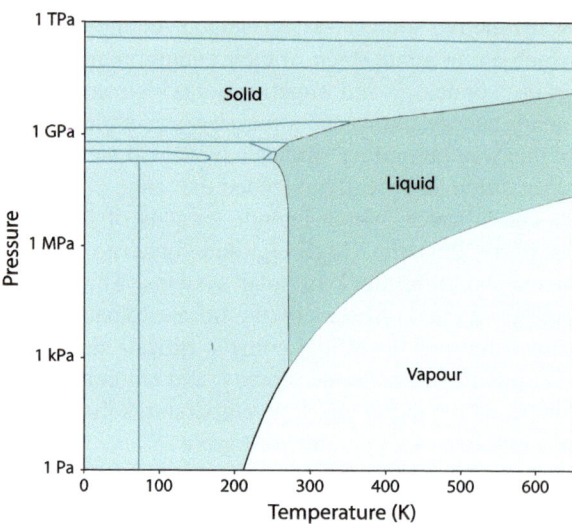

intimately associated from earliest times; gas-phase molecules froze out on dust
particles to form ices as interstellar material evolved into molecular clouds.

The term snowline defines the point beyond which water ices could exist in a
protostellar disk around a star. For the young Sun it would have been at ~2.7 AU.

Methane is thought to have formed in the interstellar medium before it was
incorporated in the nebula and then trapped in clathrates ($CH_4 \cdot 7H_2O$) as the disk
cooled, and embodied in comets, icy bodies and giant planets. Titan too may hold
primeval methane clathrate whereas on Earth it is mainly biological in origin and on
Mars it may derive from hydrothermal reactions with olivine-rich material [17, 29].

Two important present-day reservoirs for ices are the disk-shaped Kuiper-
Edgeworth Belt, just beyond the orbit of Neptune at 30–1000 AU, and the Oort
Cloud, a spherical shell which surrounds the Solar system and extends from 20,000
to 50,000–200,000 AU from the Sun (Frontispiece). The former hosts 10^8–10^9 or so
short-period comets, including Jupiter-Family comets (orbital period (p) < 20 year)
and Halley-type comets (p 20–200 year), as well as the dwarf planets Pluto, Haumea
and Makemake. The Oort cloud holds perhaps as many as 3–5.10^{12} predominantly
icy bodies and is the source of long-period comets (p > 200 year) such as Hale-Bopp.

The dimensions of the Oort Cloud and to a lesser degree the Kuiper Belt are
debated: both are hypothetical zones for which evidence has been understand-
ably slow to accumulate. The original formulation for the eponymous cloud by
J.H. Oort (following E. Öpik) in 1950 depended on orbital data for 19 comets, a
database expanded by 1978 to 300 [7]. The Kuiper Belt was first substantiated in
1992–3 when two bodies were detected at 50 AU [6]. That figure now stands at
over 1000 Kuiper Belt Objects.

Although the nuclei of comets are widely thought to contain 'the most pristine material' in the solar system it is not certain whether this protosolar interstellar dust has remained totally unmodified rather than being evaporated before

incorporation in comets [21]. But there is little doubt that Oort bodies represent icy planetesimals which formed among the giant planets in the outer part of the protoplanetary disk, that is they consist of reworked primeval material.

In order to clarify the sequence of events that led to the ubiquity of water in the solar system—in oceans, icy moons, the giant planets and cometary ice—we must turn to the D/H ratio (^{2}H or deuterium relative to H or protium), a value that combines ancestry with upbringing as it reflects environmental changes and also points to plausible evolutionary sequences.

Deuterium is a product of Big Bang nucleosynthesis, with the primordial ratio with hydrogen-1 (D/H) estimated on the basis of theoretical models as 2.61×10^{-5} (0.0000261) [33]. Modelling suggests that the conditions that prevail in protostellar disks are not conducive to the production of high levels of 'deuterated' water, as the ionization or levels that are derived from galactic cosmic rays are inadequate, and a major source must therefore be presolar, that is to say derived from interstellar sources via the molecular cloud [10]. This line of argument boils down to stating that a source of water must have been available to all candidate solar systems.

The search for a source of the Earth's water inevitably considered comets as a possible option and this became credible when astronomers were able to quantify the incidence of cometary infall from space imagery, but the isotopic composition (D/H or H_2/H) of four comets including Halley's (1P/Halley) seemed to conflict with this suggestion [2] as it is very different from the mean value for the Earth's oceans (Figs. 2.5 and 2.6). The organic load of Halley's comet [9] would

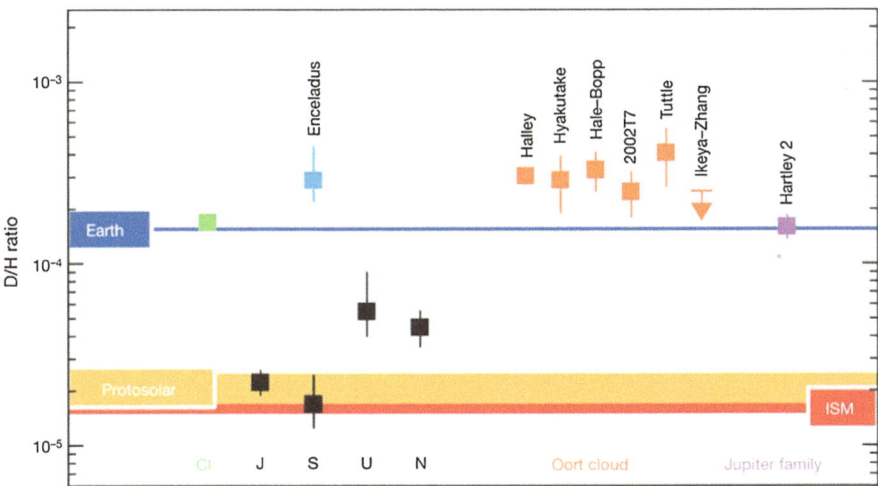

Fig. 2.5 D/H values for Oort-cloud comets (orange squares, the atmosphere of the giant planets Jupiter (J), Saturn (S), Uranus (U) and Neptune (N) (*black symbols*), water in the plume of Saturn's moon Enceladus and in CI carbonaceous chondrites (*light blue* and *green symbols*, respectively). Earth's ocean (*blue line*), protosolar value, local interstellar medium (ISM). Error bars 1σ. After Hartogh et al [15] with permission

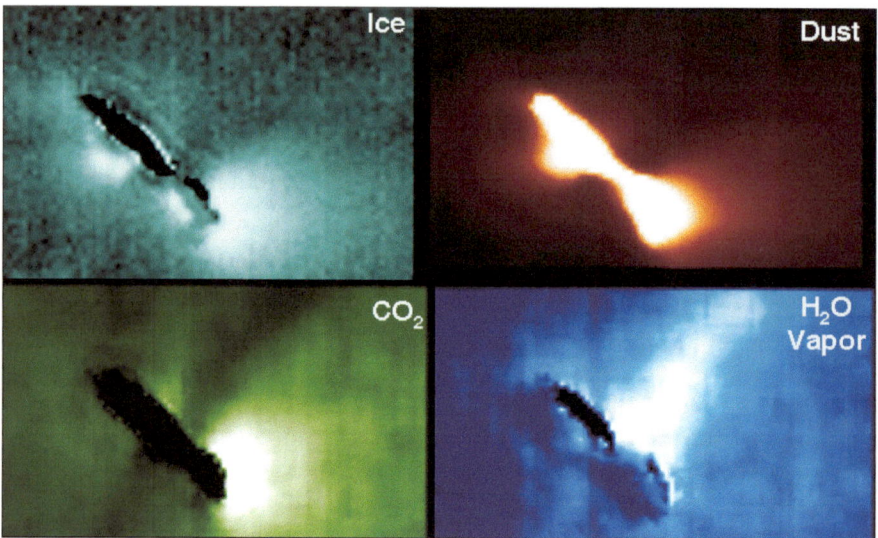

Fig. 2.6 Infrared scans of Comet 103P/Hartley 2 by NASA's EPOXI mission spacecraft show CO_2, dust and ice being ejected from one location and water vapour from another. 4 November 2010. Courtesy of NASA/JPL-Caltech/UMD

consequently be a poor guide to the likely cometary contribution to an extraterrestrial origin of life on Earth.

However, Herschel measurements on the Jupiter-family comet 103P/Hartley 2 show a D/H ratio similar to that of the Earth's oceans [2]. This complicates the issue of the provenance of Jupiter-family comets in the disk but revives the case for a cometary origin of the Earth's water rather than meteorites originating in protoplanets in the outer asteroid belt [28].

Moreover, samples from the Kuiper Belt comet 81P/Wild 2 that were returned to Earth in 2006 by the Stardust mission include some grains from other stars, but the bulk of the solids are solar system materials which formed over an extended time period at both the highest and the lowest temperatures in the early solar system, perhaps originating the hot inner regions before being transported beyond the orbit of Neptune. They include fragments of CAIs and also tungsten, molybdenum, ruthenium, and other refractory materials.

CAIs and Chondrules

Calcium-aluminium-rich inclusions (CAIs) are found in chondritic meteorites—that is to say primitive meteorites characterised by the spherical bodies known as chondrules (Fig. 2.7)—which yield some of the oldest ages for the solar system. Chondrites are a mixture of presolar and solar nebula materials and also asteroidal debris, as chondrules continued to form after early planetesials had formed and collided [32].

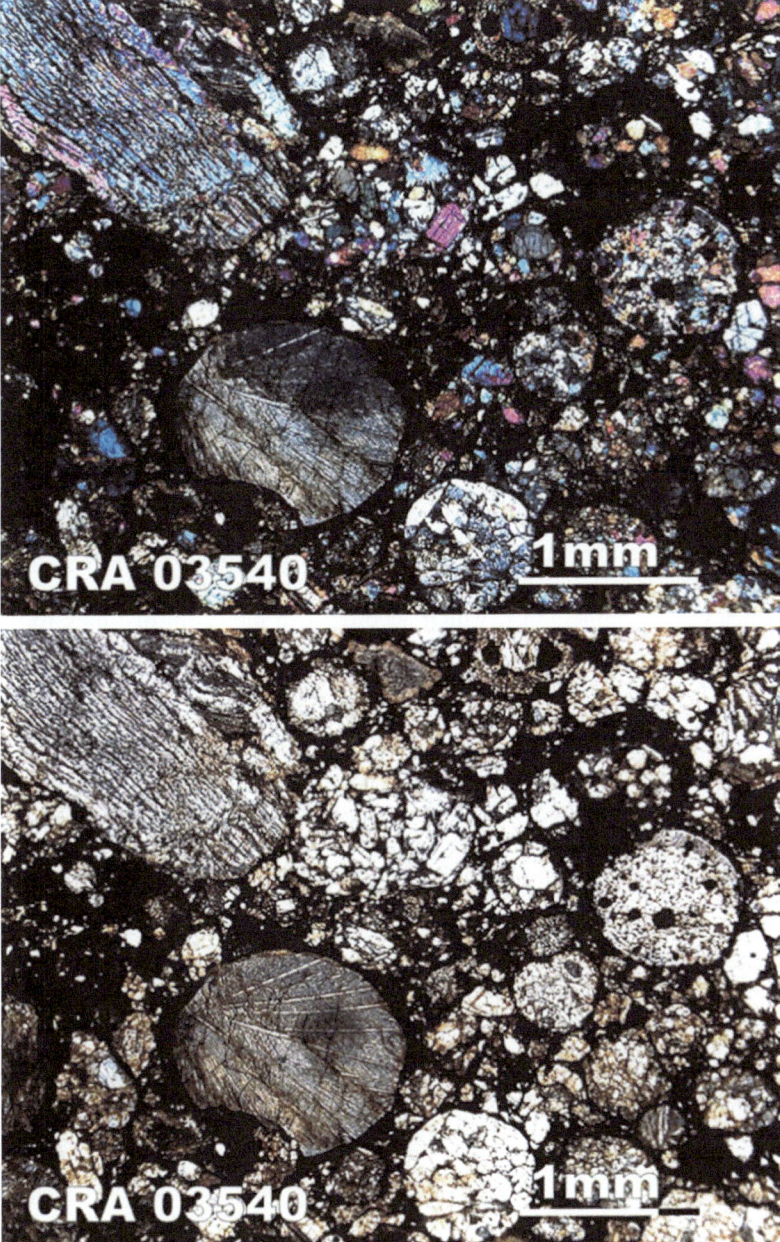

Fig. 2.7 Thin section under cross-polarised light (*upper*) and plane-polarised light (*lower*) **of the L3 chondrite** CRA 03540 **showing distinctive chondrule textures:** barred olivine type (*upper left corner*), radial pyroxene type (*lower centre*), **and** porphyritic type (near the *scale bar*). Courtesy of NASA

CAIs are not restricted to the CV chondrites where they were first reported and the basis for the original studies, the Allende meteorite, is now known to have undergone substantial postaccretion reprocessing [27]. Isotopic studies suggest that CAIs formed near the young Sun and were latter scattered to different accretion sites. They solidified from partly to completely molten material, perhaps more than once, interacted with gases or liquids, and were reheated by shock events.

Chondrules are spherical bodies typically measuring several microns (μm) and composed of olivine or pyroxene or both. They are thought to have formed from molten or partly molten material 1–3 Myr after the CAIs although their ages overlap [14]. The melting was accomplished by flash heating followed by cooling slow enough to favour crystal growth. Suggestions for the process include solar nebula lightning, sudden exposure to sunlight, collision between asteroids, shocks resulting from gravitational instability within the nebula and planetesimal bow shocks.

On balance the thermal histories indicated by the chondrules favour melting driven by gravitational instability associated with spiral arms in the solar nebula [13]. Even if the collisional mechanism prevailed and involved several generations of planetesimals primary chondrule formation it would have lasted a mere 2 Myr whereas CAI formation could have lasted several 10^5 years [24].

Some of the larger (>2 μm) dust fragments in Comet 81P/Wild 2 resemble chondrules in their mineralogy, which accords with the claim that they formed in the inner solar system; the finer material has a more complicated signature and could include dust from the outer nebula or diverse conditions in the inner nebula [30]. The comet has been in orbit beyond Neptune since it formed and, unlike some asteroids whose composition has been 'compromised' by heating and wetting, apparently retains a faithful record of early solar system conditions.

Chondrites of the CI group contain up to 20 % of water and minerals that have been altered in the presence of water (Fig. 2.8) such as hydrous phyllosilicates similar to terrestrial clays, oxidized iron in the form of magnetite, and sparsely distributed crystals of olivine scattered throughout the black matrix. In addition,

Fig. 2.8 PSD-XRD patterns for the CI chondrites Alais (*green*), Orgueil (*blue*) and Ivuna (*red*), showing the main identified phases. Alais and Ivuna are offset on the Y-axis for clarity. After King et al [20] with permission

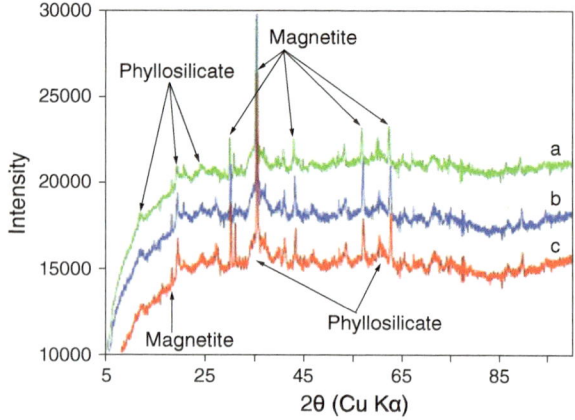

Fig. 2.9 Two bright stars illuminate a mist of PAHs in this image, a combination of data from Spitzer and the Two Micron All Sky Survey (2MASS). Courtesy of NASA/JPL-Caltech/2MASS/ SSI/University of Wisconsin-Spitzer

they contain certain amounts of organic matter like amino acids, which are of course the building blocks of life on earth.

The D/H value for water locked in carbonaceous chondrites reflects an origin in the inner early solar system [18] but, as the chondrules of CI chondrites were never heated above 50 °C during their formation, there must have been an active interplay between outward turbulent diffusion and inward advection of the water within the disk. Low values would have prevailed in the hot inner disk and higher values in the outer disk but, especially in the early stages of disk evolution, D/H values rose again in the outermost disk as water that was incorporated early in the disk's evolution was pushed out. Thus Oort cloud comets that formed early would have D/H values similar to those of the giant planets. In short, the early solar nebula was not fixed in composition, as it received fresh contributions from the molecular cloud as well as undergoing internal reorganisation [36].

PAHs

Polycyclic aromatic hydrocarbons (PAHs) appear to be abundant in the universe. For some years a mysterious set of emissions in the infrared was detected around many celestial objects in our galaxy and in other galaxies. In 1985 they were tentatively identified, at least in part, with large PAHs [2], ubiquitous component of organic matter in space (Fig. 2.9), having been identified in interstellar graphite grains, protoplanetary nebulae, circumstellar disks, IDPs comets and meteorites [34] as well as in planetary settings including the upper atmosphere of Titan. Abundant PAHs were detected on fresh fracture surfaces in Martian meteorite ALH84001, which was found in Antarctica in 1984 and displays carbonate globules and a number of isotopic and morphological features which are deemed organic in origin [25] (see Chap. 7). Cosmic ray exposure data suggest ALH84001 was in space about 16 Myr before landing 13,000 years ago, but the PAHs in ALH84001 were at concentrations 10^3–10^5 times greater than in Greenland ice dating from the last 400 year and show every indication of being 'indigenous' to the meteorite. PAHs may form when complex organic substances are exposed to high temperatures or pressures; they consist of as many as 6 fused benzene rings containing only carbon and hydrogen [3].

From being suspected of contributing to infrared emission spectra, PAHs have emerged as playing a key role in photoelectric heating of interstellar gas. And there is general acceptance that they contained ~10 % of the carbon in the interstellar medium (~40 % being in the form of dust) when the protoplanetary disk began to collapse [12] and that they first formed as early as two billion years after the Big Bang.

Laboratory studies suggest that interstellar conditions can transform PAHs into more complex organic molecules [15] by hydrogenation and oxygenation at temperatures as low as 5 K when subjected to UV radiation. This may account for the lack of PAH signatures in interstellar ice.

More important, the UV flux in protoplanetary or circumstellar environments is far higher than in dense molecular clouds. A possible route to greater complexity is via pyrimidine, also carbon-rich, which is suspected of condensing on the surfaces of cold icy grains in dense molecular clouds. If subjected to UV (see Fig 4.1) radiation in the laboratory, pyrimidine in ice rich in H_2O (as well as methane, ammonia or methanol) may yield uracil, cytosine and thymine, 'informational subunits of DNA and RNA' [31]. Nucleobases are of course found in several meteorites including the carbonaceous chondrites Murchison and Orgueil. The debate over the timing and setting of the origin of life [19] refuses to focus down.

References

1. Allamandola LJ, Tielens AGGM, Barker JR (1985) Polycyclic aromatic hydrocarbons and the unidentified infrared emission bands: auto exhaust along the Milky Way. Astrophys J 290: L25-L28
2. Altwegg K et al (2015) 67/P Churyumov-Gerasimenko, a Jupiter family comet with a high D/H ratio. Science 347, doi:10.1126/science.1261952

3. ATSDR (2015) Agency for Toxic Substances and Disease Registry at http://www.atsdr.cdc.gov/csem/
4. Bernatowicz TJ, Zinner E eds (1997) Astrophysical implications of the laboratory study of presolar materials. AIP, New York
5. Bernatowicz TJ et al (2005) Constraints on grain formation around carbon stars from laboratory studies of presolar graphite. Astrophys J 631:988-1000
6. Boice DC (1997) Kuiper Belt. In: Shirley JH, Fairbridge RW (eds) Encyclopedia of planetary sciences. Kluwer, Dordrecht, 381
7. Boice DC, Fairbridge RW (1997) Oort, Jan Hendrik (1900-1992), and Oort cloud. In: Shirley JH, Fairbridge RW (eds) Encyclopedia of planetary sciences. Kluwer, Dordrecht, 559
8. Bradley J (2010) The astromineralogy of interplanetary dust particles. Lecture Notes in Physics 815:259-276
9. Capaccioni F et al (2015)The organic-rich surface of comet 67/P Churyumov-Gerasimenko as seen by VIRTIS/Rosetta. Science 347, doi:10.1126/science.aaa0628
10. Cleeves LI et al (2014) The ancient heritage of water ice in the solar system. Science 345:1590-1593
11. Connelly JN et al (2012) The absolute chronology and thermal processing of solids in the solar protoplanetary disk. Science 338:*651-655*
12. dePater I, Lissauer JJ (2001) Planetary sciences. Cambridge Univ Press, Cambridge
13. Desch SJ, Morris MA, Connolly HC Jr and Boss AP (2012) The importance of experiments: constraints on chondrule formation models. Meteor Planet Sci 47:1139-1156
14. French B, MacPherson G, Clarke R (1990) Antarctic meteorite teaching collection. At http://curator.jsc.nasa.gov/
15. Gudipati MS and Yang R (2012) In-situ probing of radiation-induced processing of organics in astrophysical ice analogs – novel laser desorption laser ionization time-of-flight mass spectroscopic studies. Astrophys J Lett 756: doi:10.1088/2041-8205/75/1/L24
16. Hartogh P et al (2011) Ocean-like water in the Jupiter-family comet 103P/Hartley 2. Nature 478:218–220
17. Hersant F (2004) Enrichment in volatiles in the giant planets of the Solar System. Planetary and Space Science 52: 623–641.
18. Jacquet E, Robert F (2013) Water transport in protoplanetary disks and the hydrogen isotopic composition of chondrites. Icarus 223:722-732
19. Joseph R, Schild R (2010) Biological cosmology and the origins of life in the Universe. J Cosmology 5:1040-1090
20. King AJ, Schofield PF, Howard KT, Russell SS (2015) Modal mineralogy of CI and CI-like chondrites by X-ray diffraction. Geochim Cosmochim Acta 165:148-160
21. Li A and Greenberg JM (2002) In dust we trust. In: Pirronello V and Krelowski J (eds) Solid State Astrochemistry. Kluwer, Dordrecht, 1-44
22. Lodders K (2003) Solar System abundances and condensation temperatures of the elements. Astrophys. J. 591:1220-1247
23. Longair M S (1966) Our evolving universe. Cambridge University Press, Cambridge
24. Lugmair GW, Shukolyukov A (2001) Early solar system events and timescales. Meteor planet sci 36:1017-1026
25. McKay DS et al (2010) Search for past life on Mars: possible relic biogenic activity in Martian Meteorite ALH84001.Science 273:924-930
26. McKee CF and Ostriker EC (2007) Theory of star formation. Annu Rev Astron Astrophys 45:565-687
27. MacPherson GJ (2003) Calcium-aluminium-rich inclusion in chondritic meteorites. In Davis AM (ed) Treatise on Geochemistry, Elsevier, 201-246
28. Morbidelli A et al (2000) Source regions and timnescales for the delivery of water to the Earth. Meteor Planet Sci 35:1309-1320
29. Mousis O, Chassefière E, Holm N G, Charlou J -L, Rousselot P (2015) Methane clathrates in the solar system. Astrobiology 15:308-326

30. Ogliore RC et al (2015) Oxygen isotope composition of coarse- and fine-grained material from Comet 81P/Wild 2. Geochim Cosmochim Acta 166:74-91
31. Sandford SA et al (2014) Photosynthesis and photo-stability of nucleic acids in prebiotic extra-terrestrial environments. Top Curr Chem, doi:10.1007/128_2013_49
32. Scott ERD 2007 Chondrites and the protoplanetary disk. Annu Rev Earth Planet Sci 35:577-620
33. Steigman G, Romano D, Tosi M (2007) Connecting the primordial and galactic deuterium abundances. Mon Not Roy Astr Soc 378:576-580
34. Visser R et al (2007) PAH chemistry and IR emission from circumstellar disks. Astron Astrophys 466:229-241
35. Williams JP et al (2000) The structure and evolution of molecular clouds: from clumps to cores to the IMF. In: Mannings V et al (eds) Protostars and Planets IV. Univ. of Arizona Press, Tucson, 97-120
36. Yang L, Ciesla FJ, Alexander CMO (2013) The D/H ratio of water in the solar nebula during its formation and evolution. Icarus 226:256-267

Chapter 3
Assembly

Abstract The solar nebula incorporated material from its parent molecular cloud to form a protoplanetary disk and this in turn became differentiated into a protosun with an array of other bodies some rocky, some icy and some gaseous. In the standard scheme dust grains cohered to form aggregates which grew into 10–100 km planetesimals some of which benefited from runaway and disproportionate (oligarchic) growth and became embryo planets. Large embryos acquired atmospheres gravitationally, with those beyond the snowline (2–4 AU from the Sun) developing into ice and gas giants. Crossing orbits sometimes led to collisions which account for anomalous orbital geometries and the formation of rings and satellites including our Moon, and which may explain the loss of much of its mantle by Mercury. A number of recent models postulate substantial changes in the orbits of the giant planets early in solar system history, with an initial compact configuration destabilised by the 2:1 resonance crossing of Jupiter and Saturn about 700 Myr after the gas of the protoplanetary disc had dissipated.

Analysis of our solar system benefits from the fact that the birth of the central star can be dated directly rather than (as elsewhere) by stellar modelling [13]. The main source for the crucial ages is radioactive age determination of early solar system material recovered from meteorites. The values given by Wasserburg [41] are 4.563×10^9 yr and 4.576×10^9 yr rounded up to 4.6×10^9 yr, and a CAI from meteorite NWA 2364 from North Africa later gave a Pb–Pb age of 4.568×10^9 yr [5]. In a word, reassuringly consistent.

That meteorites contain material dating from the earliest days of the solar system is of course an assumption but one for which there is abundant circumstantial evidence. The oldest meteorites available for analysis are the carbonaceous chondrites (see Chap. 2) and they include a fine-grained, volatile-rich matrix thought to have formed from part of the nebula that was already depleted in volatile elements [3]. By volatile is meant condensing at <650 K and by moderately volatile condensing at c 1350–650 K. Evidence of element exchange between chondrules and matrix suggest they formed in the same (inner) part of the nebula; the nature of the episode is consistent with a shock wave. An early attempt to model the processes

© Springer International Publishing Switzerland 2016
C. Vita-Finzi, *A History of the Solar System*, DOI 10.1007/978-3-319-33850-7_3

operating in the presolar nebula which was based on analysis of the Allende carbonaceous chondrite (which fell in Mexico in 1969) likewise saw a supernova remnant as trigger which pushed into an interstellar cloud [40].

Mixing through migration, both among the planets and their satellites and the half million or so asteroids that have now been documented, [15] is now seen as a key feature—in the assembly of our solar system. Migration is a term used for a change in the orbit of a planet or satellite resulting from its interaction with planetesimals, another planet or the disk itself. The interaction may be mediated by orbital resonance.

Migration features prominently in the Nice model of solar system evolution (named after the Observatoire de la Côte d'Azur where the model was first crafted) in which Jupiter, Saturn, Uranus and Neptune formed more closely spaced and nearer the Sun than now until there was a 2:1 resonance between Jupiter's orbit and that of Saturn. (Early concern with the orbits of Jupiter and Saturn was mentioned in Chap. 1.) As a consequence, Saturn was forced outwards into the early Kuiper Belt and the asteroid belt.

The resulting disturbance is blamed for the Late Heavy Bombardment (LHB) when the terrestrial planets experienced intensive impact cratering [17]. Traces of the heaviest phase of asteroid attack are thought to include the lunar maria, the Caloris basin on Mercury and the large craters in the southern hemisphere of Mars.

Preliminary analysis of impact melt sheets had endorsed the notion of a 'cataclysmic bombardment' by large planetesimals that had affected the Earth, the Moon and possibly the entire inner solar system about 3.85 ± 0.1 Gyr ago. Dating of melt breccias at Apollo 16 sites by ^{40}Ar–^{39}Ar coupled with petrographic analysis later showed that there were at least four impact episodes within a mere 200 Myr or so during the period between 3.96 and 3.75 Gyr [33], an interesting refinement on the original model. Some authorities now extend the bombardment on the basis of beds of spherules formed from molten rock to 3.47–1.7 Gyr [22]; others believe that, as indicated by partial ocean evaporation, giant asteroid impacts occurred as late as 3.29–3.23 Gyr [27].

The Nice model explains a peak in impact rate about 3.9 Gyr following changes in the orbits of the Giant Planets. It appears to account for much else. For instance there are about 1000 of the asteroids known as Jupiter Trojans (named after the large asteroids Achilles, Hector and Agamemnon) in two large swarms with a 1:1 resonance with Jupiter. Their orbit was predicted in terms of the three-body problem, where a small body moves under the gravitational influence of the Sun and a planet, hence their location at 'Lagrange points' where this relationship provides a stable orbit and is the location of choice for artificial satellites such as SOHO which wish to keep the Sun under surveillance. When it was that the Jupiter Trojans took up their position, however, is uncertain [30]. The Nice school suggests that the Jupiter-Saturn resonant interaction first rendered the Trojans' orbits chaotic and then nudged them into their present co-orbital motion with Jupiter [32].

Table 3.1   The Titius-Bode scheme

	x	T-B (AU)	Distance (AU)
Mercury	0	0.4	0.39
Venus	3	0.7	0.72
Earth	6	1.0	1.00
Mars	12	1.6	1.52
Ceres	24	2.8	2.77
Jupiter	48	5.2	5.20
Saturn	96	10.0	9.54
Uranus	192	19.6	19.2
Neptune	388	38.8	30.06
Pluto	256	77.2	39.44

The Titius-Bode 'Law', first formulated by JE Bode in 1778, states that the spacing of the solar system planets corresponds to a simple mathematical progression: add 4 to x in the table and divide by 10. The answers give the distances in AU and are successful as far as Neptune. The table highlighted a gap that came to be filled by the asteroid belt

In short, the Nice framework would seem to combine the stability that accompanies resonance effects with the confusion that may be triggered by the migration of large solar system bodies.

The much derided Titius-Bode scheme seemed to endorse a stable pattern (Table 3.1). It indicated a gap in the sequence from Mercury to Saturn where Ceres, then treated as a planet but since demoted to the status of dwarf planet, was eventually discovered. As Bode noted in 1772 'Can one believe that the Founder of the universe had left this space empty?' [20]. The Titius-Bode law predicted the location of Uranus but the calculated orbital radius was greatly in error, as it was for Pluto. The formula has proved successful with regard to the satellites of Uranus [29]. It has been dismissed as merely the outcome of scale invariance and rotational symmetry in the protoplanetary disk [18], but this surely qualifies Titius-Bode as deterministic rather than as the product of chance.

But it is futile to seek a static model when the present pattern is the product of prolonged history. Not that a dynamic view of the relationships between solar system bodies is revolutionary: growth, capture, collision and other kinds of departure from orbital rigidity have always been implicit in analyses of planetary observation. The need remains to understand how individual bodies are formed before analysing their interrelationships in space and time.

Accretion

In some molecular clouds part of the constituent dust and gas in due course becomes concentrated in a core amounting to some 10^4 solar masses ($M_\odot$). The core attracts further material gravitationally until it collapses into clumps of

Fig. 3.1 The Orion Nebula, Messier 42 Courtesy of NASA, ESA, M Robberto (Space Telescope Science Institute/ESA), the Hubble Space Telescope Orion Treasury Project Team and L Ricci (ESO)). Six protoplanetary disks or proplyds are marked

10–50 $M_{\odot}$. According to numerical models the collapsing core may develop into a disk in ~10,000 yr and the disk will have a lifetime of less than 1–10 Myr [43]. There is also some evidence that protoplanetary disks are found only around stars less than 10 Myr old and are a part of star formation.

The notion of protostars bordered by circumstellar disks has gained fresh favour from direct observation of extrasolar planets and of young stars [25], notably by optical imaging by the Hubble Space Telescope (Fig. 3.1), [43] while the evolution from disk to protostar can be traced on systems at different stages of development thanks to analysis of temperature and rotation by infrared measurement, which as we saw earlier can penetrate the obscuring dust. Thus it has taken until the late 20th century for the insights of Swedenborg, Kant and Laplace to be validated instrumentally.

The gravitational collapse model was challenged by Alfvén and Arrhenius [1]. They claimed that no observation had yet been found in its support and they preferred a hetegonic process, a term they coined for the formation of secondary bodies around primary bodies. The key item in support of their hypothesis was spin isochronism: with a few exceptions which could be explained by resonance or tidal effects (as with Mercury and Venus; Mars remained for them a puzzle), they claimed that most bodies ranging in size between 10^{18} and 10^{30} g have a rotational period within a factor of two of 8 h, whereupon they argued that a theory to explain planetary formation around the Sun should apply to the formation of satellites in general.

The rotational period argument is questionable but in any case a more important obstacle faced by Laplace's nebular theory and its variants is that the planets hold 0.14 % of the mass of the solar system yet exhibit 99.5 % of its angular momentum, with Jupiter contributing 60 % of the total. This is mainly a consequence of their large orbital planetary radii: in the crucial formula for rotational angular momentum, $Mr^2\omega$, where M is the mass of the rotating body, r its distance from the axis of rotation and ω its angular velocity, the value r is squared. The problem does not arise in planet-satellite systems where the ratio between the angular momentum of the satellite and of the central body is very small [45].

The issue has been tackled in different ways. One approach is to invoke the redistribution of mass (and hence angular momentum) in the disk, a ruse which appears to be sanctioned by the vigorous solar winds exhibited by T tauri stars and thus presumably the protoSun. In the prototype of this youthful category as the gas cloud condenses under gravity some of the infalling gas, heated by collisions, is ejected in jets (Fig. 3.2) and with material that contributes to the solar mass.

The redistribution of material in the nebula is consistent with the proposal by Percival Lowell [28] that the mass of the nebula is at the centre, in his view because the nebula was formed by collision between two bodies. Other early models include that by See [6, 36], who argued that the planets formed in the nebula independently and were then captured by the Sun. This amounts to arguing that the mismatch in their momentum potential is not the outcome of a common origin in the disk. Taking the capture notion further, it has been suggested that numerous rogue planets

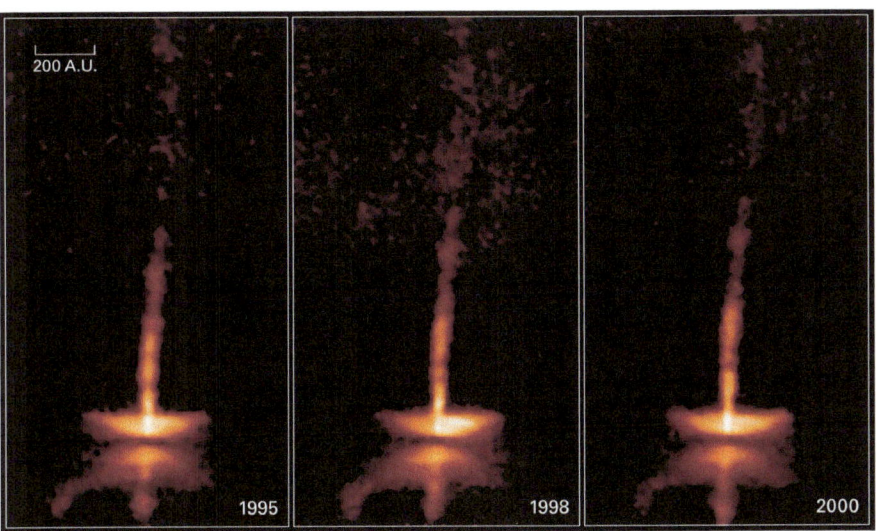

Fig. 3.2 Changes over 5 yr period in the dust disk and jets of protostellar object HH30, a newborn star about half a million years old. Diameter of disk is about 450 AU. Imaged with the Wide Field and Planetary Camera 2 aboard NASA's Hubble Space Telescope. Courtesy of NASA, Alan Watson (Universidad Autónoma de México), Karl Stapelfeld (JPL), John Krist and Chris Burrows (ESA/Space Telescope Science Institute)

discarded by their parent star systems have been seized by other stars, and this is held to account for the location of some planets orbiting far from their host (presumably 'foster') stars and for the existence of some double-planet systems [34].

At a more modest level, planetesimal capture by young stars is often invoked at the runaway stage. By then of course the angular momentum problem will have been resolved; indeed, it would appear that most of the disk material surrounding stars older than c. 5 Myr will have been converted into planetesimals or planets, accreted onto the star or dumped from the system [39].

Whatever one's views about the hetegonic thesis, some kind of accretion is evidently central to most accounts of solar system origins [11]. Calculations based on plausible dust density in a turbulent medium and on the assumption that all collisions result in sticking—perhaps through magnetic, mechanical or electrostatic forces or through partial fusion through impact—suggest that after 4000 yr particles with a mass of 10 g might be common and that after 10,000 yr much of the dust might form part of bodies some of them measuring thousands of kilometers [24] and qualifying as protoplanets.

The planetesimal theory was initially formulated by Chamberlin [10] in contradistinction to nebular and 'meteoroidal' models [6]. 'Nebular' alludes to 'Laplacian and other gaseous hypotheses', 'meteoroidal' mainly to the ideas of Darwin [14], who had invoked 'the aggregation of meteorites' as a means of forming planetary bodies. Chamberlin's principal point was that in existing models aggregation was driven by gravity whereas he saw it as the outcome of crossing orbits. Ironically the planetesimal thesis is now integrated into the ruling Laplacian scheme as shown in summary statements of the standard model of planet formation: [23] 'turbulent clumping' within the disk concentrates particles into substantial entities (>1 km) intermediate in size between small bodies which cohere by chemical bonds and van der Waal (intermolecular) forces and large bodies for which gravity promotes growth [12].

Once there is a substantial number of planetesimals—so the conventional view runs—and gravitational attraction between pairs of planetesimals comes to dominate accretion we enter a phase of runaway growth. It culminates in the emergence of planetary embryos within 10^5–10^6 yr coexisting with the remaining planetesimals— hence the term oligarchic growth. The remaining gas damps any eccentric behaviour by protoplanets but once the gas is depleted (in 1–10 Myr) gravitational effects render the system unstable leading to orbital crossing or collision between protoplanets [22].

In theory the chemical composition of the planetesimals should reflect the temperature gradient to be expected with increasing distance from the star subject, of course, to variations in the timing of accretion and to disturbance by turbulence. Being promoted by loss of gas one might expect planetesimal formation to begin in the hotter, inner disk. This in turn might favour heating by radioactive elements in the mix, such as aluminium 26 (^{26}Al), and thus more sticking.

Beyond the critical radius we have met as the snowline, at between 2 and 4 AU, temperatures would have been low enough for water ice to condense and for the accretion of hydrogen and helium to form the giant planets. The requisite ice or rock cores, originating in collision between planetesimals or gravitational clumping [4], can then serve as nuclei for gas giants such as Jupiter and Saturn, which are composed mainly of hydrogen and helium, and ice giants such as Neptune and

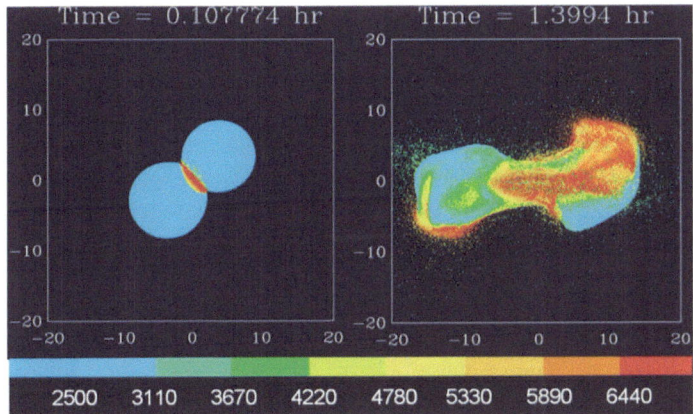

Fig. 3.3 Initial stages of collision between two protoplanets with 4–5 × the mass of Mars. Computer simulation [7]. Such simulations allow a wide range of masses, impact angles and speeds to be postulated Temperatures in K. Image courtesy of NASA and Southwest Research Institute

Uranus which are composed of heavier elements. The cores had to become massive enough to attract and hold sufficient gas gravitationally and this had to happen before the gas in the protoplanetary disk had dissipated [26].

A number of simulations for the maturation of planetary embryos to terrestrial planets in the inner disk yield timescales of 30–100 Myr. The relationship of these estimates to ages obtained by direct dating of Earth materials is not straightforward, especially when we consider the possibility, raised in one discussion of terrestrial age based on radiogenic noble gases in the atmosphere, that the result (4.45 ± 0.02 Gyr) represents the time when the last substantial planetesimal struck the Earth and in so doing 'reinitialized' the global clocks [46].

Modelling (Fig. 3.3) suggests that such collisions are commonplace. They do not inevitably result in perfect accretion, however, hence the alternative category of 'hit-and-run', which has been invoked to resolve such puzzles as Mercury's retention of volatiles when (as is often posited) it was stripped of much of its mantle by one or more impacts. Indeed the process could be part of a sequence in which Mars and Mercury were the last two survivors of 20 or so embryos which contributed to the accretion of Venus and Earth [2].

Satellites and Rings

The solar system contains six satellite systems yoked to a planet (Earth, Mars, Jupiter, Saturn, Uranus and Neptune) or a dwarf planet (Pluto, Eris and Haumea) to a total, by February 2016, of 173. Jupiter is currently known to have 67 satellites of which 59 qualify as irregular in having large orbits, eccentricities and inclination. All Jupiter's satellites are in its Hill sphere, but, as we saw, it also hosts two Trojan clouds at Lagrange points L4 and L5. Minor planets have a total of over 200 satellites, and about 119 bodies in the asteroid belt also have satellites, as do 79 trans-Neptunian objects. Mercury and Venus are thus anomalously bereft.

Rocky material dominates in the satellites of the inner solar system whereas frozen volatiles, including water ice, ammonia, methane, nitrogen, carbon monoxide, carbon dioxide and sulphur dioxide are major components of those of the outer solar system. Many of them show evidence of geological activity after their formation, sometimes driven by tidal heating or the consequences of impact. The resulting differentiation is discussed in Chap. 5.

The prevailing model has long been that each of the giant planets developed a miniature solar system in its accretion disk, with prograde, near-circular orbits. The capture of passing planetesimals could provide additional satellites with irregular orbits which are prograde or retrograde according to where the capture occurred. Satellites are termed irregular if they have large orbits, eccentricities and inclinations.

Capture could plausibly occur if a planetesimal was slowed down by the planet's atmosphere ('gas-drag'), through the three-body effect considered briefly above, by collision, or through 'pull-down' capture. The last process, which appears to be confined to Jupiter by virtue of its magisterial proportions, arises if there is a sudden spurt in the planet's mass with a consequent increase in its Hill sphere, the zone where a planet's gravitational pull over its satellites is dominant. Capture is considered an early solar system phenomenon because the opportunities for such activity are nowadays rare [21].

The difficulties that confront students of what could be called planetary palaerodynamics are brought home by considering the Moon, whose origin remains in dispute despite decades of theoretical discussion, wide-ranging remote sensing, and direct inspection, sampling and laboratory analysis. It has been argued [44] that most of the gain in understanding gained from the Apollo missions came from advances in dynamical analysis and a better grasp of the planetary context, but much of the credit 'probably goes to digital computers.'

The many models that have been proposed for the Moon's origin [44] can be grouped into those that invoke the gravitational capture by the Earth of an existing body; co-evolution with the Earth; and the aggregation of material such as meteorites, planetesimals, or debris created by collision with the Earth or by the disintegration of a large planetesimal by tidal forces originating in the Earth.

There are many obstacles to a single irrefutable explanation. In coaccretion the Earth acquires a disk of solid particles from which the Moon is then formed. Coevolution is of course implicit in condensation within a Laplacian nebula; in a sense it is also exemplified in theories invoking fission, as in the suggestion by Darwin [14] that the Moon was ejected when the Sun's semidiurnal tide was in resonance with the free oscillations characterising the viscous Earth + Moon spheroid, a view which remained in vogue, with the Pacific Ocean basin as the resulting scar, until it was invalidated on geophysical arguments and, recently, by the recognition of the geological youth of the Pacific basin.

Coaccretion was long the majority favourite until it was shown on several counts that it was inconsistent with the angular momentum of the Earth-Moon system, whereupon collisional ejection was seen the most plausible of the alternatives

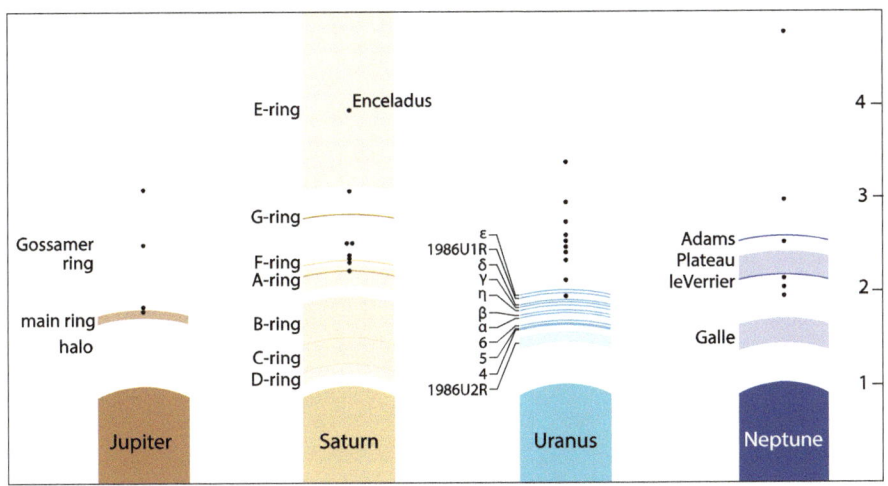

Fig. 3.4 Planetary disks, rings and major satellites of the giant planets. The rings lie within the Roche limits of each planet except for G and E rings of Saturn (see text). Scale represents distance in planetary radii for each of the planets. After A D Fortes in Vita-Finzi, Fortes AD (2013) Planetary Geology (2nd ed) Dunedin, Edinburgh, with permission

[19, 44], and it was refined into impact by a Mars-sized body which generated a disk of debris from which the Moon coalesced some 4.5 Gyr (Gigayear or 10^9 years) ago [7]. This thesis explains why the Moon is poor in iron, as the core from the impactor would have merged with the Earth, and it gives the Earth the right initial spin rate to result in today's. But the strong similarity in the stable isotopic (^{16}O ^{17}O and ^{18}O) composition of the Moon and the Earth [42] is difficult to explain unless both impactor and proto-Earth were about 50 % the mass of the Earth so that the collision destroyed the impactor, deformed the Earth and produced a disk composed of material from both bodies. Alternatively, both bodies were of very similar composition from the outset, [9] and modelling suggests this is the case for many planet-impactor pair involving a giant impactor [31].

This discussion brings us back to the giant planets (Fig. 3.4). Tidal factors also dominate models for the origin of the rings and satellites of Saturn. The rings of Saturn consist of 90–95 % water ice whereas an ice-rock mixture might be expected. The Roche limit, the radial distance from a primary within which a large satellite is disrupted by tidal forces, was invoked by Édouard Roche himself in 1849 to explain why only rings and small moons are found close to the planet [35]. When applied to complex, dynamic material which is also subject to magnetic forces, gas drag and meteoritic and cometary impacts it can only be an approximate value: thus for a satellite with the density of water is at 2.1 R_s (the equatorial radius of Saturn, 60,330 km) and thus lies within the A ring. In any case, rather than disrupt an existing satellite the effect may be to prevent the aggregation of existing debris.

The Saturnian rings include particles of <1 cm diameter (the majority being >1 m) which would be slowed down by solar radiation and the exosphere

Fig. 3.5 Image of Jupiter on the NASA Infrared Telescope Facility, Mauna Kea, Hawaii, taken on 21 July 1994, showing Io (*top right*), the Great Red Spot (*lower left*) and six of the Shoemaker-Levy 9 impact sites, with the brightest due to Fragment R. Courtesy of NASA IRTF Comet Science Team

and are presumably being replenished. Numerical modelling favours as the source of the fragments the tidal removal of the outer icy layers from a Titan-sized satellite [8], although a contribution may sometimes come from cometary fragments after large events, such as the collision of Comet Shoemaker-Levy-9 with Jupiter in 1994 (Fig. 3.5) whence the ripples in these rings and in those of Jupiter [37].

These two modes may be complementary, a fusion that bears on the age of the ring system. Whereas data from the Voyagers and the Hubble telescope pointed to origin in a violent event, such as the collision of a comet with a moon, perhaps 100,000 yr ago, the data gathered by the Cassini mission indicate extensive, rapid recycling consistent with a ring system which is continually renewed but probably ancient and which includes rings (of which the Voyager spacecraft counted 1000) varying widely in age and history [16]. The same may hold for the ring systems that surround the other giant planets.

Frozen volatiles make up the bulk of the Kuiper-Edgeworth Belt, which hosts the dwarf planets Pluto, Haumea and Makemake. Like the much smaller asteroid belt it is somewhat dismissively viewed as material left over from the formation of the solar system. The Oort Cloud also consists of reworked primeval material. As noted earlier the interaction of these vast systems with the giant planets is cogently narrated by the Nice scheme and it is to be hoped that the accumulation of the ice reservoirs themselves will be clarified by current space missions and numerical models.

References

1. Alfvén H, Arrhenius G (1976) Evolution of the solar system. NASA, Washington DC
2. Asphaug E, Reufer A (2014) Mercury and other iron-rich planetary bodies as relics of inefficient accretion. Nature Geosci 7: 564-568
3. Bland PA et al (2005) Volartile fractionation in the early solar system and chondrule /matrix complementarity. Proc Nat Acad Sci USA 102:13755-13760

4. Boss AP (2003) Rapid formation of outer giant planets by disk instability. Astrophys J 599: 577-581
5. Bouvier A, Wadha M (2010) The age of the solar system redefined by the oldest Pb-Pb age of a meteoritic inclusion. Nature Geosci 3:637-641
6. Brush SG (1996) Fruitful encounters. Cambridge Univ Press, Cambridge
7. Canup RM (2004) Dynamics of lunar formation. Annu Rev Astron Astrophys 42: 441-475
8. Canup RM (2010) Origin of Saturn's rings and inner moons by mass removal from a lost Titan-sized satellite. Nature 468: 943-946
9. Canup RM (2013) Planetary science: lunar conspiracies. Nature 504: 27-30
10. Chamberlin TC, Moulton ER (1909) The development of the planetesimal hypothesis. Science 30: 642-645
11. Chambers J (2010) Terrestrial planet formation. In Seager S (ed) Exoplanets, Univ Arizona, Tucson, 1-23
12. Chiang E, Youdin AN (2010) Forming planetesimals in solar and extrasolar nebulae. Annu Rev Earth Planet Sci 38: 493-522
13. Christensen-Dalsgaard J (2009) The Sun as a fundamental calibrator of stellar evolution. In Mamajek EE, Soderblom DR, Wyse RFG (eds) The ages of stars, Proc IAU Symp 258, IAU, Cambridge Univ Press, Cambridge
14. Darwin GH (1879) On the precession of a viscous spheroid, and on the remote history of the Earth. Phil Trans Roy Soc Lond 170:447-538
15. De Meo FE, Carry B (2014) Solar system evolution from compositional mapping of the asteroid belt. Nature 505: 629-634
16. Esposito L (2014) Planetary rings: a post-equinox view (2nd ed). Cambridge Univ Press, Cambridge
17. Gomes R, Levison HF, Tsiganis K, Morbidelli A (2005) Origin of the cataclysmic Late Heavy Bombardment period of the terrestrial planets. Nature 435: 466-469
18. Graner F, Dubrulle B (1994) Tititus-Bode laws in the solar system, Part I: scale invariance explains everything. Astron Astrophys 282: 262-268
19. Hartmann WK (2014) The giant impact hypothesis: past, present (and future?) Phil Trans Roy Soc A 372: 20130249. doi:10.1098/rsta.2013.0249
20. Jaki SL (1972) The early history of the Titius-Bode law. Am J Phys 40:1014-1023
21. Jewitt D, Haghighipour N (2007) Irregular satellites of the planets: products of capture in the early Solar System. Annu Rev Astron Astrophys 45: 261-295
22. Johnson VC, Melosh HJ (2012) Impact spherules as a record of an acient heavy bombardment of Earth. Nature 485:75-77
23. Kokubo E, Ida S (2012) Dynamics and accretion of planetesimals. Prog Theor Exper Phys. doi:10.1093/ptep/pts032
24. Lewis JS (1995) Physics and chemistry of the solar system (rev ed). Academic, San Diego
25. Lissauer JJ (2006) Planet formation, protoplanetary disks and debris disks. In Armus L, Reach WT (eds) The Spitzer space telescope, ASP Conf 357:31–38
26. Lissauer JJ, Stevenson DJ (2007) Formation of giant planets. In Reipurth B et al (eds) Protostars and planets V, Univ Arizona, Tucson, 591-606
27. Lowe DR, Byerly GR (2015) Geologic record of partial ocean evaporation triggered by giant asteroid impacts, 3.29-3.23 billion years ago. Geology 43:535-538
28. Lowell P (1903) The solar system. Houghton Mifflin, Boston
29. Lynch P (2003) On the significance of the Titius-Bode law for the distribution of the planets. Mon Not R Astron Soc 341: 1174-1178
30. Marzari F et al (2002) Origin and evolution of Trojan Asteroids. In Bottke WF Jr et al (eds) Asteroids III, Tucson, Univ Ariz Press, 725-738
31. Mastrobuono-Battisti A, Perets HB, Raymond SN (2015) A primordial origin for the compositional
32. Morbidelli A et al (2005) Chaotic capture of Jupiter's Trojan asteroids in thre early solar syste,. Nature 435:462-465

33. Norman MD, Duncan RA, Huard JJ (2006) Identifying impact events within the lunar cataclysm from ^{40}Ar-^{39}Ar ages and compositions of Apollo 16 impact melt rocks. Geochim Cosmochim Acta 70:6032-6049

34. Perets HB, Kouwenhoven MBN (2012) On the origin of planets at very wide orbits from the recapture of free floating planets. Astrophys J 750: 83. doi:10.1088/0004-673x/750/1/83

35. Roche E (1849) La figure d'une masse fluide soumise à l'attraction d'un point éloigné. Mém Ac Sci Montpellier 1: 243-262

36. See TJJ (1910) Researches on the evolution of the stellar systems II. The capture theory of cosmical evolution. Nichols, Lynn MA

37. Showalter MR, Hedman MM, Bjurns JA (2011) The impact of Comet Shoemaker-Levy-9 sends ripples through the rings of Jupiter. Science 322: 711-713

38. Spohn T, Johnson T, Breuer D (eds) (2014) Encyclopedia of the solar system (3rd ed). Elsevier, Paris

39. Van der Marel N, et al (2015) Resolved gas cavities in transitional disks inferred from CO isotopologs with ALMA. Astr Astrophys 585: A58. doi:10.10651/00004-6361/201526988

40. Wark DA (1978) Birth of the presolar nebula: the sequence of condensation revealed in the Allende meteorite. Astrophys Space Sci 65:275-295

41. Wasserburg GJ (1995) Solar models with helium and heavy-element diffusion. Rev Mod Phys 67:781-808

42. Wiechert U et al (2001) Oxygen isotopes and the moon-forming giant impact. Science 294:345-348

43. Williams JP, Cieza LA (2010) Protoplanetary disks and their evolution. Annu Rev Astron Astrophys 49:67-117

44. Wood JA (1986) Moon over Mauna Loa. In Hartmann WK, Phillips RJ, Taylor GJ (eds) Origin of the Moon , Lunar Planet Inst, Houston TX, 17-55

45. Woolfson MM (2011) On the origin of planets. Imperial Coll Press, London

46. Zhang Y (1998) The young age of Earth. Geochim Cosmochim Acta 62: 3185-3189

Chapter 4
The Solar Nucleus

Abstract The Sun, which embodies much of the mass of our solar system, has evolved substantially since its coagulation from the nebula. It is at the heart of the solar system in many ways besides its obvious gravitational role, notably through the radiation it emits at a range of wavelengths, the magnetic heliosphere and the solar wind. All fluctuate periodically, irregularly and cumulatively in response to factors working within the Sun or externally, notably orbital cycles, interaction between bodies, and interstellar matter.

The great misfortune of the astrologers, claimed Voltaire [22], is that the sky has changed since the rules of the art were established. The Sun, which at the equinox was in Aries in the time of the Argonauts, is to-day in Taurus; and the astrologers, to the great ill-fortune of their art, to-day attribute to one house of the Sun what belongs visibly to another.

Astrologers are not alone in founding their predictions on shifting sands, as are satirists in mocking philosophers: all definitions for our solar system are constantly changing both in their limits and in their properties. The Sun itself, which embodies 99.9 % of the mass of the solar system, has evolved substantially since its coagulation from the nebula, and is by no means the sedate, benevolent figure of infantile art. Gravity, the Newtonian criterion, leaves no imprint that is currently mappable, but roundabout reasoning sometimes allows us to reconstruct the new orbit of one or other solar system body through capture or mischance which helps to redefine the boundaries of the system or at least show that it has undergone significant changes. Geological evidence allows us to trace changes in the solar wind and in the interplanetary magnetic field back for millions of years though very discontinuously.

The most thoroughly investigated component of current solar behaviour is its output of radiation. Total solar irradiance (TSI), the oldest measure of solar activity and long misleadingly termed the solar constant, would seem the simplest measure of present and past solar influence. The solar flux is an average of $1366 \, \text{W m}^{-2}$ at 1 AU at wavelengths with the bulk in the visible range (Fig. 4.1); at times of high solar activity the range extends to X-ray wavelengths at 1 nm or less.

© Springer International Publishing Switzerland 2016

C. Vita-Finzi, *A History of the Solar System*, DOI 10.1007/978-3-319-33850-7_4

Fig. 4.1 The electromagnetic radiation (EMR) spectrum. Wavelengths in metric units and for radio and TV also the corresponding frequencies in Hertz. Note position of visible and UV bands. *nm* nanometre = 10^{-9} m

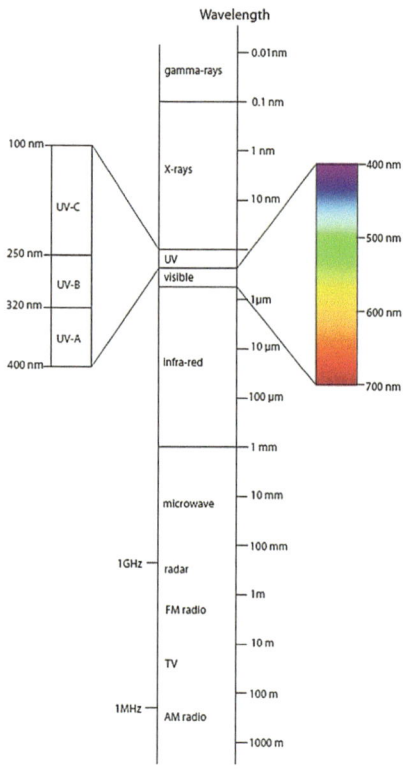

As a power output solar luminosity can be expressed as a total of 3.85×10^{26} W or $1 \, L_{\odot}$. Its measurement presented grave problems until balloons, rockets and spacecraft could raise the key devices above the Earth's atmosphere. In view of the predilection for certain wavelengths by the eyes of different organisms and by the pigments that facilitate photosynthesis [6] it would be of great interest to evolutionary biologists to know whether, when and how far the solar spectrum has varied over the aeons. For the aeons prior to the launch of the 7-ERB instrument on the NIMBUS satellite in 1978 we have to rely mainly on indirect measures, notably the generation of cosmogenic isotopes in the atmosphere by cosmic rays and deposited in ice caps and tree rings. As the cosmic ray flux is blocked by the solar wind, the isotopes it generates provide an inverse measure of the Sun's activity; unfortunately the flux is also influenced by the interplanetary and solar magnetic fields.

Nevertheless highly informative estimates of wavelength-dependent changes in solar flux have been made for the range between 0.1 nm and 160 μm, that is in the visible and infrared range. The estimates used present-day solar data from 0.6 Gyr after the Sun's zero age on the Main Sequence (see Fig. 4.2) to 2.3 Gyr beyond its present age of 4.6 Gyr. The Main Sequence (MS) is the name given to the dominant grouping of the nearest stars on a scattergram of luminosity (or absolute

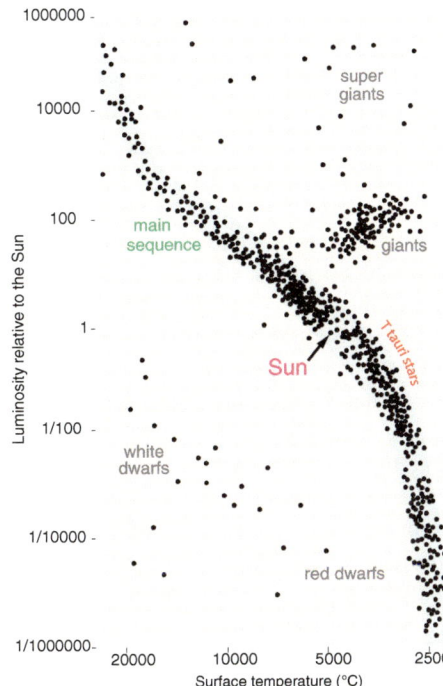

Fig. 4.2 A Hertzsprung-Russell diagram, a summary plot of the brightest stars seen from Earth in terms of their luminosity (or intrinsic brightness) and surface temperature. Note location of the Sun near the middle of the main sequence, log scale of *y* axis and temperature (*x* axis) increase from right to left

magnitude) against colour (or temperature). As originally (and independently) devised by Ejnar Hertzsprung and Henry Norris Russell [9, 15] it used spectral class for the x axis (which traditionally increases from right to left) and visual magnitude for the y axis; the version shown here is a summary of a version based on the 10,000 brightest stars as seen from the Earth. The Sun is near the mid-point of the MS, which is occupied by stars that fuse hydrogen in their cores. They leave the MS once they have exhausted the hydrogen in their core.

The solar body at the outset of its MS career, that is at zero age in the main sequence or ZAMS, is taken to consist of homogeneous gases in hydrostatic equilibrium, and in devising SSMs the hydrogen/helium (H/He) ratio is adjusted so that the Sun's luminosity at its present (meteorite-derived) age of 4.6 Gyr is reproduced by the model. In one such calculation the data consist of solar mass; solar radius from observation; luminosity from the solar constant (determined instrumentally from satellites); the composition of the solar photosphere (or visible surface) as indicated by spectral analysis and sometimes expressed simply as the ratio between the abundance of elements heavier than helium and that of hydrogen; and solar age, often simplified to 4.6×10^9 years [2].

By confining the analysis to solar flux rather than the amount of radiation received by the Earth the problems of atmospheric history and variation across the solar system could be shelved [3]. To put it another way, the findings are of value throughout the solar system.

The results were validated by comparison with two solar-type stars, K^1 Cet and EK Dra. Numerical modelling and analogy with other stars also allows time calibration of nebular history early in solar system history. The time required to form the nebular disk, for example, may have taken less than 10^3 years [19]. In models which view the Orion Nebula as a plausible nursery, and place T tauri stars at the heart of any protoplanetary disks that nucleate in it (see Chap. 3), the solar nebula displayed radially a thermal and gravitational gradient which governed the chemistry of the solids and ices that condensed at different distances from the young Sun and on either side of the snowline. The time taken for the protoSun to get started may in fact have been as little as 10,000 year; [8] protostars can move to the Main Sequence after some 50 Myr and, if similar in mass to our Sun, stay on it for 10^{10} year. For the Sun the ensuing history is founded on some version of the mathematical standard solar model (SSM), whereby hydrogen in the Sun's interior is progressively transformed to helium, with a corresponding increase in the Sun's core temperature and thus in its luminosity.

Through this or some such comparable route we can arrive at the initial helium abundance, which is a key influence on luminosity and a crucial value in the study of the chemical evolution of galaxies, and to an assessment of progressive changes in the Sun's influence on other bodies in the solar system.

Gravity

The collapse of a gas and dust cloud into planetary cores is driven by gravity and resisted by gas pressure; if sufficient, gravity may ultimately trigger nuclear fusion and propel the infant star into the ranks of the protostars. The gravitational force is of course not uniformly distributed, and it contributes to the onset and maintenance of the rotation that governs subsequent events. The luminosity of the star is proportional to the energy expended in counteracting gravity [8]. It has been conjectured that the annihilation of dark matter, estimated to be 5–10 times more abundant than

Fig. 4.3 Kepler's First Law of planetary motion states that the orbit of ever planet is an ellipse with the Sun at one of its two foci, (*a* and *b*, or *a* and *c*); the Second Law that a line joining the Sun and the planet sweeps out equal areas during equal intervals of time (A_1, A_2), hence more rapid orbital motion near the Sun. The total orbit times for planet 1 and planet 2 have a ratio $ab^{3/2}:ac^{3/2}$

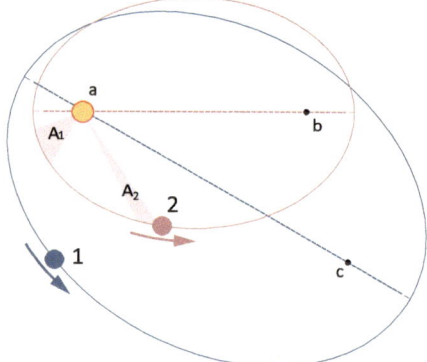

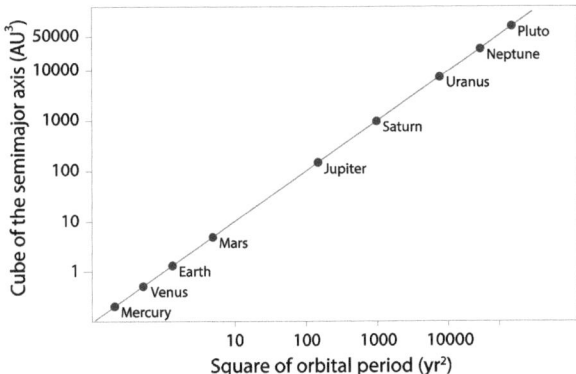

Fig. 4.4 A plot of the square of the planet's orbital period against the cube of its semi-major axis for the planets and Pluto to demonstrate Kepler's Third Law

visible matter on the basis of its gravitational impact [12], may have powered the earliest giant stars (measuring 1 AU or more) instead of nuclear fusion [18].

Kepler's laws are shown in summary fashion in Figs. 4.3 and 4.4. The present solar mass $M\odot$ can be computed from Kepler's third law as redefined by Newton to take account of his law of gravitation, to wit $P^2 = a^3/M$ where P is the planet's orbital period, a is the semi-major axis of the planet; to simplify the calculation we can assume the Earth's orbit is circular so that a is one AU; and M is the mass of the star compared with which the Earth's is insignificant. The result is $M\odot = {\sim}2 \times 10^{30}$ kg. As shown below in Chap. 7, various stratagems have been employed to reconcile the evidence for relatively warm temperatures on early Mars and Earth with a key prediction of the H/He mechanism, as a Sun fainter by 30 % than that of today should have gazed down on a frozen Earth whereas geology indicates liquid water at the latest by 3.8 Gyr ago. Solutions to this 'Faint Young Sun paradox' include postulating an initial mass for the Sun 0.01–0.07 greater than today's.

Whatever the validity of the 1.07 $M\odot$ estimate, annual mass loss now due to the solar wind is about 3×10^{-14} M. The p-p chain is repeated c. 9.2×10^{37} s^{-1} and leads to a loss of 4.4×10^9 kg sec^{-1} so that, at this (the current) rate, the Sun over 4.5 Gyr ago would have been more massive by a mere 1/10,000 of the present value. Indeed, although the Sun loses 10^9 tonnes/sec, direct measurement by the Michelson Doppler Imaging instrument on the SOHO satellite of solar diameter over much of a solar cycle has found no evidence that annual diameter change exceeds 15 mas (milliarcsecond or 1/1000 of a second of arc) [11]. Of course mass loss may have been greater in the past especially if the Sun experienced a spell as a T tauri star. As it happens, the presence of neon-21 (^{21}Ne) in a number of carbonaceous chondrites (including the Murchison meteorite) indicates irradiation by energetic protons from the young Sun.

The mass loss required for a warmer young Sun is consistent with observational data from young MS stars such as π^{01} Uma as well as κ^1 Cet and β Com, high-precision calculations from helioseismology, and estimates for solar wind activity derived from evidence of solar irradiation in the lunar soil or regolith (Fig. 4.5) and radiation damage [5]. The present solar wind points to a mass loss over

Fig. 4.5 Astronaut Alan L
Bean samples lunar regolith
during Apollo 12 mission.
Courtesy of NASA

4.5 Gyr of a mere $\sim 10^{-4}$ $M\odot$; the noble gas isotopes in the lunar regolith suggest
that the solar wind flux has not changed over the past several billion years [23]
whereas the abundance and isotopic composition of nitrogen in the lunar soil suggest that the intensity of the solar wind over the last billion years, if not lunar history in its entirety, was 3 times greater on average than today [4].

Moreover, modelling of the plasma dynamics from the wind base up to 3 AU
which refers to the Sun at 0.7 Gyr, at 2 Gyr and at the present day (4.65 Gyr), and
among other things takes the strength of the Sun's magnetic field into account [1],
suggests that the solar wind was twice as fast, 50 × denser and twice as hot than
now at 1 AU at 0.7 Gyr. This provisional finding will doubtless colour discussions
of conditions on other young stars as well as the erosion of the atmospheres of the
Earth and the other terrestrial planets and of their habitability in terms of the flux
of GCR (that is to say cosmic rays that originate outside the solar system though
probably from elsewhere in the galaxy) and ozone shielding against aggressive
solar radiation.

Equally problematic are attempts to assess the validity of the SSM through
the neutrino count, as, although it refers to the present day, gamma ray photons
generated by the nuclear fusion processes at the Sun's core are thought to take
many thousands of years to reach the solar surface because the random walk they
experience is coloured by the mean free path which among other things depends
on variable conditions in the solar interior. Neutrinos are generated in large numbers by thermonuclear reactions including fusion in the Sun's core (Fig. 4.6) at
an estimated rate of 10^{38}/s with energies that depend on which element is fused.
The number recorded between 1968 and 2002 in detectors using perchloroethyl-
ene, water or gallium persistently remained between 30 and 54 % of the amounts
predicted by the Standard Solar Model (SSM) at the accepted core temperature, a
temperature that was confirmed in 2001 by helioseismology.

In 1998, work at the Super-Kamiokande observatory in Japan showed that neutrinos generated by the interaction between cosmic rays and the Earth's atmosphere

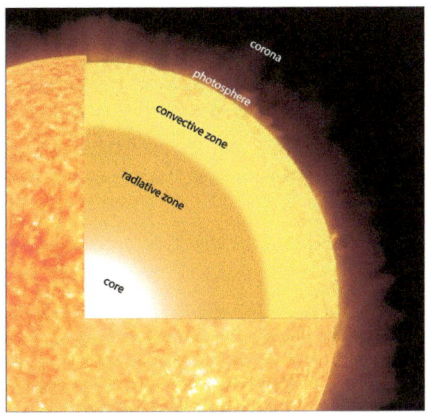

Fig. 4.6 Main subdivisions of the solar interior. The boundary between the radiative zone and the convective zone is the tachocline; the base of the solar atmosphere is occupied by the chromosphere

appear to change flavour on the way to the detector, and measurements in 2001 at the Sudbury Neutrino Laboratory in Canada revealed that the metamorphosis also applies to solar ('electron') neutrinos. The discovery of the flow of particles, misleadingly termed rays, by Victor Hess in 1911–1912 depended on a gold leaf electroscope carried aloft in a balloon. In fact Hess detected the secondary particles produced by the interaction of CRs with atmospheric nuclei, but he succeeded in showing the source was from space rather than (as he himself originally suspected) from the Earth.

One of Hess's flights was made during a solar eclipse so that the Sun could be ruled out as the source of the CRs, but the Sun enters the story indirectly as the variable source of the solar wind that deflects GCRs. Solar cosmic rays (SCRs), also known as solar energetic particles (SEPs), are associated with flares. They too are dominated by protons but at lower energies than GCRs.

More important than the book-balancing was the revelation that, contrary to the assumptions made in the SSM, neutrinos have mass, which allows them to change flavour and thus restores the neutrino flux to that predicted by the fusion model.

The Heliosphere

Sometimes described as the region of space dominated by the solar wind, the heliosphere is intimately linked to the solar corona. The existence of a solar wind was hypothesised by Parker [13] and had been inferred from the homely yet persuasive observation by Ludwig Biermann in 1951 that the tails of comets (we now know to be the ionised gas of the two tails) whatever their orbits, invariably point away from the Sun, and in due course calculations showed that the corona would release a continuous stream of particles.

The reality of a solar wind was first convincingly demonstrated by measurements by Luna 1 in 1959 and by the Mariner 2 Venus probe in 1962. It consists of a plasma which is released by the solar corona and which is composed mainly of electrons and protons with a minor component of alpha particles. Its density [16] is about 5 cm^{-3} and it embodies magnetic field which is twisted into a spiral by the Sun's rotation. The heliosphere has rightly been termed magnetic field inflated by the solar wind.

Today the solar wind is thoroughly monitored, and its speed and composition as recorded by the NASA Advanced Composition Explorer (ACE) satellite at the L1 libration point between Earth and Sun. It travels at between 260 and >750 km s^{-1} with an average of about 400 km s^{-1} at a rate of about 1 m tons s^{-1}, and appears to emanate mainly from the Sun's coronal 'holes' near its poles.

The area swept by the wind took even longer to define. It was by observing the relative thrust of GCRs and the solar wind that NASA's Voyager I and Voyager II, launched in 1977, were able to observe the boundary of the heliosphere. (Note that the predictive power of Newtonian physics allowed planning the Voyager missions to take advantage of a planetary alignment that occurs only every 176 years.) Voyager I crossed the termination shock and entered the heliosheath in August 2004. The heliopause was first signalled by an increase in CR hits on the craft's instruments by about 25 % after January 2009 and by 5 % in a week beginning on May 7 of the same year. In December 2011 Voyager 1 detected a region where the solar system's magnetic field had doubled in strength in response to pressure from interstellar space. The solar wind's energetic particles had declined almost by a half whereas high energy electrons from outside the heliosphere increased 80-fold. In August 2012 it crossed the heliopause 121 AU (128×10^9 km) from the Sun and entered interstellar space.

Voyager II had taken a different route and did not leave the solar wind embrace until 30 August 2007 at a point nearly 10^9 km closer to the Sun than Voyager I, thus showing that, as suspected, the interstellar magnetic field deforms the heliosphere by driving it closer to the Sun.

Data from the two Voyagers, the Interstellar Boundary Explorer (IBEX) spacecraft and the work of computer modellers show that, instead of the single comet-like tail it was thought to display owing to its collision with the interstellar magnetic field, the heliosphere is dominated by two jets one above the north and one below the south poles of the Sun, resulting in a crescent form. The magnetic fields indicated by IBEX are consistent with the pattern of GCR flux observed on Earth.

The record of cosmogenic isotopes such as radiocarbon (^{14}C) and beryllium 10 (^{10}Be) that can be recovered from ice cores and tree rings is a guide to the former strength of the heliospheric shield and hence of the impact of the solar wind. Primary GCRs consist mainly of high energy protons; their interaction with the atmosphere or the Earth yields secondary GCRs. Since their discovery in 1912, cosmic rays have acquired (it might be better said retained thanks to the association of 'rays' with the malign death rays of science fiction) a reputation for potentially damaging DNA and thus opposing the habitability of planets that are insufficiently protected against energetic GCRs by a deep atmosphere or a strong magnetic field.

The GCRs interact with atmospheric molecules to generate a range of cosmogenic nuclides. They include carbon 14 (^{14}C) and beryllium 10 (^{10}Be), which are then stored in tree rings, marine sediments and stratified ice [21]. The ^{10}Be record, currently the longest, goes back 800,000 years in the EPICA Dome C core in Antarctica, but the contribution to its modulation by the Earth's magnetic field and distortion of the evidence by climatic change hamper attempts to reconstruct changes in the heliosphere.

That changes in the Sun had an impact on the Earth's magnetic field was first accepted, despite Lord Kelvin's scepticism, when the large flare observed by Richard Carrington in 1859 was followed by disruption of telegraph systems within one or two days, indicating particles ('solar corpuscular radiation') presumably travelling at 10^3 km s^{-1} [14]. The precise nature of this radiation could not be grasped before the necessary advances in theoretical and observational physics as well as spectroscopic measurements of the temperature of the Sun's corona during 1939–1942.

All these advances still did not define a volume in space dominated by the solar wind. The answer hinged on an observation made before the solar wind had been identified: that, as the number of sunspots rose, there was a decline in the intensity of cosmic rays. This effect discovered by Scott Forbush in 1937 and named after him [7], results from blockage of CRs by magnetic fields in the solar wind, the very effect that allowed Voyager I's managers to recognise the satellite's exit from the heliosphere.

That the heliosphere fluctuates in volume in the short term is a consequence of changes in solar wind flux and thus of the processes that govern the corona. A cyclic association between the solar cycle and cosmic ray flux had been detected by van Allen [20], among. others, but any secular or at any rate long-term effect could not be investigated before an explicit measure of coronal activity had been formulated. A coronal index CI has been proposed for this purpose using the averaged daily irradiance indicated by ground observation of part of the corona (the green coronal line at 530.3 nm) and calibrated using data from the SOHO spacecraft [17]. The results over 5 solar (sunspot) cycles (1953–2008) were found to be highly correlated with CR data from neutron monitors, albeit with lags between the two measures of up to 410 days. In other words the association between cosmic rays and solar activity is now securely demonstrated, which means, among other things, that any reliance on the solar shield against cosmic rays, something that concerns aircraft crew as well as astronauts, [10] has to take into account its variability at time scales measured in seconds and days, as well the familiar 11 year solar cycle, position in the heliomagnetic field and long-term trends.

References

1. Airapetian VS, Usmanov AV (2016) Reconstructing the solar wind from its early history to current epoch. Astrophys J 817:L24
2. Christensen-Dalsgaard J (2008) ASTEC – the Aarhus STellar Evolution Code. Astrophys Space Sci 316: 13–24

3. Claire MW et al (2012) The evolution of solar flux from 0.1 nm to160 μm: quantitative estimates for planetary studies. Astrophys J 757: 95
4. Clayton RN, Thiemens MH (1980) Lunar nitrogen:evidence for secular change in the solar wind. In: O\Ppepin RO, Eddy JA, Merrill RB (eds) The ancient Sun. Pergamon, New York,463-473
5. Crozaz G et al (1977) The record of solar and galactic radiations in the ancient lunar regolith and their implications for the early history of the Sun and Moon. Phil Trans Roy Soc Lond A 285:587-592
6. Ensminger PA (2001) Life under the Sun. Yale Univ Press, New Haven
7. Forbush SE (1937) On diurnal variation in cosmic ray intensity. Jour Geophys Res 42:1-16
8. Gargaud M et al (2012) Young Sun, early Earth and the origins of life. Springer, Berlin
9. Hertzsprung E (1911) Uber die Verwendung photographischer effektiver Wellenlängen zur Bestimmung von Farbenäquivalenten. Publ Astrophys Observ Potsdam 22, 63
10. Kerr RA (2013) Radiation will make astronauts' trips to Mars even riskier. Science 340:1031
11. Kuhn JR et al (2004) On the constancy of the solar diameter II. Astrophys J 613:1241-1252
12. Ostriker JP, Mitton S (2013) Heart of darkness. Princeton Univ Press, Princeton
13. Parker EN (1958) Dynamics of the interplanetary gas and magnetic fields, Astrophys J 128:664-676
14. Parker EN (2007) Conversations on electric and magnetic fields in the cosmos. Princeton Univ Press, Princeton
15. Russell HN (1913) 'Giant' and 'dwarf' stars. Obs 36:324-329
16. Russell CT (2001) Solar wind and interplanetary magnetic field: a tutorial. In Song P, Singer HJ, Siscoe GL (eds) Space Weather, Am Geophys Un, Washington,73-89
17. Rybanský M (2001) Rušin V, Minarovjech M (2001) Coronal index of solar activity - solar-terrestrial research. Space Sci Rev 95: 227-234
18. Spolyar D, Frese K, Gondolo P (2008) Dark Matter and the first stars: a new phase of stellar evolution. Phys Rev Lett 100:051101
19. Taylor SR (2001) Solar system evolution: a new perspective. Cambridge Univ Press, Cambridge
20. Van Allen JA (2000) On the modulation of galactic cosmic ray intensity during solar activity cycles 19, 20, 21, 22 and early 23. Geophys Res Lett 27: 2453-2456
21. Vita-Finzi C (2013) Solar history. Springer, Dordrecht
22. Voltaire 1734 Philosophical Dictionary.
23. Wieler R, Heber VS (2003) Noble gas isotopes on the moon. Space Sci Rev 106:197-210

Chapter 5
Differentiation

Abstract As they evolved and interacted, solar system bodies (including the Sun) underwent changes to their interiors and surfaces. The consequences can amount to zonation into core, mantle and crust, the development of subsurface water bodies, the generation of a magnetic field, and the emergence of an atmosphere. Some of these developments prove ephemeral whether spontaneously, in response to solar evolution, or through impact with comets, planetesimals or other planets. In a sense planetary formation itself amounts to differentiation especially in the giant planets where, by accreting gas, planetesimals are transformed into cores.

Differentiation in a celestial body has been defined as the segregation of zones of different chemical or mineralogical properties [40] usually on the basis of their density. The classic outcome is the core-mantle-crust trilogy of some rocky planets (Fig. 5.1) or, in the Jovian planets or some icy moons, a silicate core encased in frozen volatiles (Fig. 5.2). Where the process is driven by heat the source may be radioactivity, potential energy liberated by accretion, tidal friction, or impact, but density segregation can be achieved without heating where flow facilitates density sorting.

In general, differentiation is a path to stability, but it can promote volcanic activity or even have orbital consequences. The outcome of differentiation is evidently more likely to persist in rocky bodies than in icy or gaseous ones but it may be interrupted or reversed by collision or other disturbances. In some settings the avoidance of differentiation can prove advantageous: for example, carbonaceous asteroids and cometary nuclei, which as noted in Chap. 2 are considered to be among the most primitive bodies in the solar system, have survived because they are small enough to radiate away any energy resulting from accretion and have consequently remained cold enough to avoid recrystallization [11].

Differentiation can be exceedingly rapid. In the infant solar system, where gravity was the main agent for dust accretion, bodies as large as Mars took only a few million years to form. Isotopic evidence on martian meteorites (which sampled different parts of the planet) suggests that formation of a core and crust and degassing of the mantle

© Springer International Publishing Switzerland 2016   49
C. Vita-Finzi, *A History of the Solar System*, DOI 10.1007/978-3-319-33850-7_5

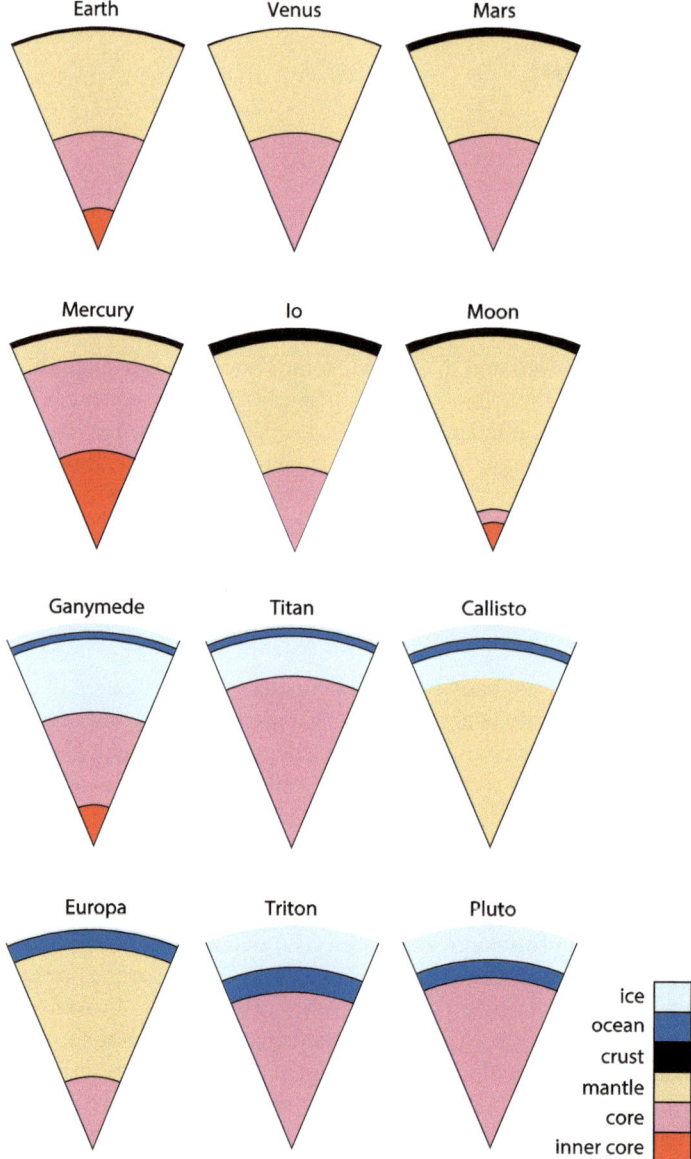

Fig. 5.1 Idealized interiors of the terrestrial planets, Pluto and selected icy satellites. In the icy bodies (lower 6) comparatively minor differences in history, composition, and location have produced a wide range of internal structures and geological processes in the larger satellites. Note that a wide range of diameters have been drawn to the same size [12]

all happened in 15 Myr or less. The fact that these zones were not subsequently 'dehomogenised' has been taken to indicate little or no convection in the Martian interior [30] but convection is implicit in evidence for early heat loss [35].

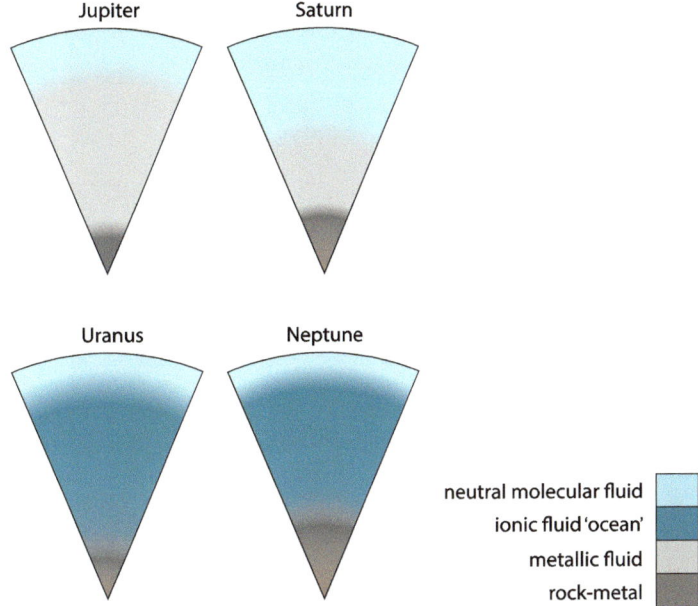

Fig. 5.2 Idealized interior of the gas giant planets. Jupiter is close to the Sun in composition; Saturn has lost a greater proportion of its He than Jupiter; Uranus and Neptune appear to incorporate ice and rock from the primordial nebula but less H and He—although of course proportionately far more than the terrestrial planets [12]

One particular isotopic clock (^{182}Hf-^{182}W) suggests that the Earth's accretion and core formation were complete c 30 Myr after the origin of the solar system [23]. Dating the differentiation into silicate and metal—essentially mantle and core—exploits a scheme formalized by the mineralogist Victor Goldschmidt [16] which among other things distinguished between lithophile elements, which combine readily with oxygen and are therefore strongly linked to silica in the Earth's crust, and siderophile, which dissolve in iron and therefore tend to migrate into the Earth's core. Hafnium (Hf) is lithophile and is retained by the silicate portions of differentiated bodies; tungsten (W) is moderately siderophile and is substantially removed from the silicates. If applied to lunar samples the Hf-W method points to a major moon-forming impact ~30–50 Myr after the solar system had formed by an impactor with a mass 1/9 that of the proto Earth [21, 23].

A very similar chronology emerges from isotopic analysis of two stony meteorites from the asteroid 4Vesta (Fig. 5.3), which suggests that differentiation was accomplished within the first 5–15 Myr of the solar system's history, whereas in the Earth-Moon system (thanks to giant impacts) differentiation continued for an additional 20 Myr [25]. The larger solar system bodies carried the process yet further by virtue of their greater mass, internal temperature and gravitational interaction with other bodies.

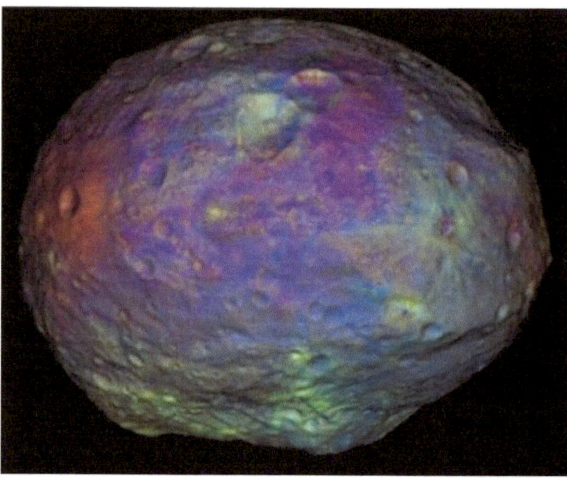

Fig. 5.3 Vesta, the largest asteroid in the Solar System (mean diameter 525 km) and the second most massive body in the asteroid belt after the dwarf planet Ceres. It was orbited by NASA's spacecraft Dawn in 2011 and 2012. Image shows varied composition indicating differentiation consistent with its status as a protoplanet. Courtesy of NASA/JPL-CALTECH/UCLA/MPS/IDA/PSI

A common source of internal heating, the radioactive decay of the isotopes aluminium-26 and iron-60 (^{26}Al and ^{60}Fe, half-lives 7.3×10^5 yr and 2.62×10^6 yr respectively), imposes a limit on the autonomous activity of the accreted body which hinges largely on its content of the two isotopes and therefore its volume. For the Moon and Mercury this amounts to about 10^9 years [11] after which the magmatic activity that can be discerned on their surfaces came to an end. These dead planets, like their active companions, will of course continue to receive the energy of impacts, which may lead to localised melting from the impact to the point where major episodes of crustal displacement result [33].

Planetary crusts come in many varieties and dimensions, as do cores and the solid and liquid mantles that encircle them. Their definition may be thermal, chemical, dynamic or some combination of attributes. The Sun's outer zones are thus generally defined by the mode of heat transfer (radiative, convective) and the Earth's vary according to the application of the analysis (seismic, tectonic, petrological). The presence of subsurface oceans, of great interest in the search for extraterrestrial life, may be inferred from gravitational data derived from spacecraft behaviour, from the escape of geysers or from distinctive surface features.

The question remains how much a body's differentiation is primarily or wholly owed to gravity (and thus density) acting benignly over time and how much to impact. Collision between embryos and planetesimals liberates high energies and may lead to the formation of a magma ocean [34]. If it is correct to assume that late stages of planetary accretion witness few, large collisions, these may well include hit-and-run-events, 'perfect mergers', and smashes [10] which retain two altered bodies.

In magma oceans that form at the surface there takes place segregation between silicates and metals leading, at least in the terrestrial planets and probably in several stages, to the formation of a metallic core, and radiometric dating shows that metal-silicate separation in the Solar System as a whole dates from the first 30 Myr of its history [45].

In contrast with the terrestrial planets, giant planets are thought to grow outwards by accreting nebular gas around a core composed of planetesimals, always providing that all the gas in the nebula has not been appropriated or has dissipated, a suggestion that appears valid for Jupiter-mass planets around solar-mass stars [24]. The process could be favoured in the outer nebula beyond the snowline because the planetesimal contribution to cores is hereabouts supplemented by ices and the parent star has too little gravitational influence to hamper growth.

The zone intermediate between crust and core—the mantle in Earth's nomenclature—is depleted in metals to the benefit of the core and will generally be essentially rocky. This is the conclusion reached about Mars in the light of martian meteorites recovered on Earth coupled with modelling studies, [46] with every indication that the martian interior was dynamically active early in Mars' history until heat transport fell below a critical level for active tectonics as well as dynamo activity.

The crust, which is the outer, solid shell of most rocky and icy solar bodies, may make up a minor percentage of its volume (1 % for Earth) but is disproportionately valuable as a source of historical information through its morphology, palaeontology, petrology and so forth and for any samples it yields to landers and as meteorites, and for what it reveals about internal processes, current dynamics, impact history and life. On Earth there are two main crustal types, the heterogeneous continental variety averaging a thickness of 35 km and the more homogeneous basaltic oceanic crust some 6 km thick. The oldest continental mineral is a zircon from a granite dating from 4.4 Gyr ago. The oldest oceanic basement basalts are ~270 Myr old [31].

The relative proportions over time of these two categories (oceanic rocks currently cover 60 % of the planet but only make up 30 % of the crustal volume) is influenced among other things by the operation of plate tectonics, the mechanism by which major fragments of lithosphere—an open sandwich of crust and upper mantle—are displaced over the lower mantle and locally thrust (or subducted) into it.

The question of when the process began bears among other things on models for the origin of life which centre on submarine hot springs located in rifts, such as the Galápagos, that separate diverging plates and, whereas the requisite data are lacking for the first 550 Myr of the Earth's history, field evidence from Western Australia reflects a rigid lithosphere subject to subduction by 4.4 Gyr ago, and various other lines of evidence point to active though localised plate tectonics at least as far back as ~3.1 Gyr ago [5, 7]. On Mars, magnetic lineations (Fig. 5.4), offsets across the Valles Marineris, and evidence of extension and compression in the Tharsis Rise are thought to be consistent with plate interaction [6, 22] in a thin (10–50 km) mainly basaltic crust over a silicate mantle which in turn overlies the iron-rich core that resulted from differentiation. The Galilean satellite Europa also displays fractures which are interpreted as displacement between blocks, and although the mantle is again composed of silicates the crust is of water ice [37].

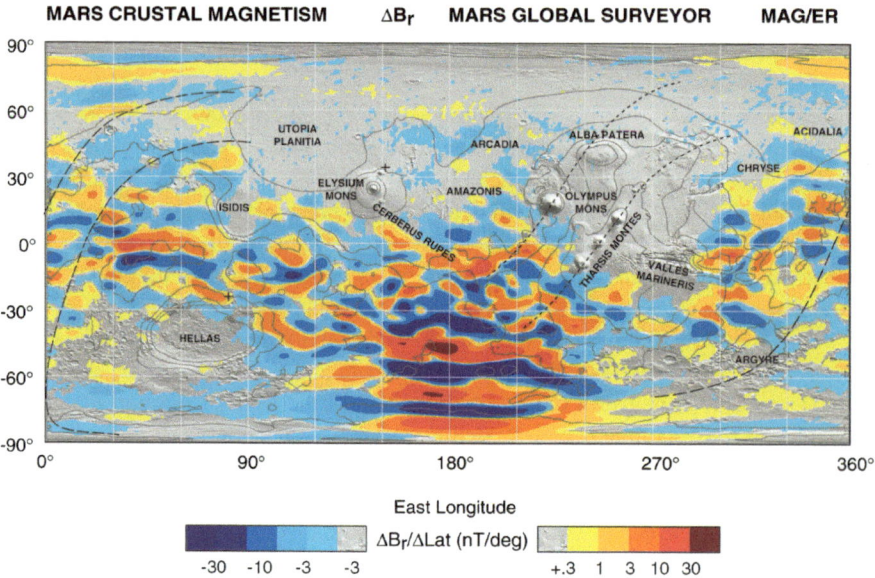

Fig. 5.4 Mars' magnetic field, with red and blue stripes representing opposite polarities. Data from Mars Global Surveyor (MGS) superimposed on Mars Observer Laser Altimeter (MOLA) map . Courtesy of NASA after Connerney JEP et al (2005) Proc Nat Acad Sci USA 102:14970-14975

Mineral Evolution

Thermal history bears on mineral history. Some of the minerals, as we saw in Chap. 2, are presolar. Others were cooked and re-equilibrated in the inner disk, although there is no evidence that the nebula was hot enough for wholesale evaporation of the major rock-forming minerals outside the orbit of Mercury [26].

The terms 'primordial' and 'evolved' are sometimes used [11] to classify components of the solar system, with the corollary that only small bodies are likely to retain primordial (i.e. solar nebula) constituents because the temperature in their interiors as a function of radioactivity will not rise sufficiently for significant differentiation to take place. The prominence of impacts in them evolution of bodies of all sizes rules out undue emphasis on radioactivity as the main engine of differentiation.

Moreover there is growing support for an alternative narrative [18]. The proposed sequence is subdivided into ten stages. Mineralogical diversity is here defined as 'micro- or macroscopic solid phases (>1 μm diameter) at crustal depths less than ~3 km' where direct interaction with microbial life is likely, as most of the 4300 or so known mineral species may owe something—if indirectly—to biochemical processes. At the start of the first era the dust particles in the molecular cloud contained about a dozen refractory oxides, silicates, nitrides and carbides; heating in the solar nebula and the associated condensation and recrystallisation gave rise to some 60 primary mineral phases incorporated in chondrules and CAIs, and further phases up to a total of about 250—'the mineralogical raw materials for Earth and other terrestrial planets'—were added by accretion, melting differentiation and impact effects. On Earth purely physical and chemical processes during the second era brought the numbers to perhaps 4500, while the third, between about 3.85 and 3.6 Gyr is termed [18] one of coevolution of the geosphere and biosphere, resulting in over 12,000 mineral novelties. The threefold scheme is applicable throughout the solar system subject to variations in timing, mechanisms and raw materials.

The accretion of planetesimals was a major driver of mineral evolution as it promoted alteration by heat generated by radioactive isotopes, T tauri solar wind, and collision, and it facilitated interaction with water. Differentiation of the planetesimals themselves, a foretaste of planetary development, is manifested in crusts and cores, the former represented by achondrites, which resemble terrestrial basalts, and the latter represented by iron-nickel meteorites. Chondritic meteorites are considered to be undifferentiated, hence their value as specimen books of unaltered primeval material.

Atmospheres and Oceans

The products of differentiation are not confined to crust, mantle and core: they include atmospheres and liquid bodies including Titan's ethane/methane oceans Some atmospheres came to be early in the body's evolution. Others reflect cyanobacterial intervention, and following the loss of primeval hydrogen and helium, are often linked to input from volcanic gases. But these gases are today composed mainly of water (95 %), CO_2 (1 %) and SO_2 (2 %). Alfvén and Arrhenius [2] argued against the 'vague notion that the Earth had somehow already formed when differentiation took place and the ocean and atmosphere began to develop'. In their view differentiation owed much to the very processes that led to the formation of the Earth so that, for example, the noble gas composition of the atmosphere resembles that of a number of meteorites and suggests that both originated in the primeval nebula, whereas the noble gas component that outgassed from the Earth's interior is dominated by radiogenic species [23]. Besides great progress in data gathering and analytical procedures their argument now suffers from evidence that atmospheric noble gases are recycled into the mantle by subduction [20].

Modelling allows us to infer which volatiles would be released when chondritic material was heated during accretion of the terrestrial planets. They include water, H_2, CO, CO_2 in proportions determined by the type of chondrite, and sulphur, P, Cl, F, sodium and potassium when temperatures of 1500–2500 K prevail [36]. The resulting atmospheres are termed primary. They have been stripped away and replaced by secondary atmospheres on all the planets bar the Gas Giants although, while Uranus and Neptune have helium abundances close those of the solar nebula, on Jupiter and Saturn some of the helium that had been in solution in metallic hydrogen in their deep interiors migrated as droplets to their cores as they cooled [29].

The smaller planetary bodies lost their primary atmospheres in various ways: thermal escape, which even now leads to the loss of several kilos of hydrogen per second from the Earth's atmosphere, interaction with the solar wind especially when there is no protective magnetic field and when the Sun was in its aggressive T tauri phase, and, especially in early Solar System history, erosion by large impacts [1] even if it was partly balanced by the volatile-rich solids and ices conveyed by the missiles. Thus measurements by the NASA rover *Spirit* on rocks at least 3.7 Gyr old in the martian crater Gusev show nickel levels five times those on martian meteorites 180 my-1.4 Ga old recovered on Earth, which suggests that the atmosphere on Mars was formerly rich in oxygen [43].

On Earth, and perhaps elsewhere, a key source of change is life. It was this that encouraged James Lovelock, author of the Gaia hypothesis, to state that one could investigate the presence of living organisms on Mars without going there by analysing its atmosphere [19]. The diagnostic gases are oxygen (20 %) and methane (2 ppm i.e. 0.0002 %).

Our terrestrial atmosphere may be almost entirely secondary and not only because the Earth harbours life. The source of the nitrogen that makes up 80 % of the Earth's atmosphere is a pertinent puzzle. If we compare the ratio between the resilient isotope neon-20 and nitrogen levels on Earth and on the Sun (i.e. as a measure of the primaeval gas) we find that nitrogen on Earth is 10,000 times more plentiful than it should be. Some argue that the excess was derived from the breakdown of ammonia (NH_3) but that merely defers the explanation. Others suggest that an inert nature allowed it to accumulate gradually whereas other, more reactive gases were incorporated in the rocks or dissolved.

A recent proposal is that the source was after all volcanic but only in the relatively oxidised conditions near converging plate boundaries where nitrogen in fluids is found predominantly as N_2 and is therefore easily degassed: elsewhere in the Earth and in the mantles of Venus and Mars, so calculations suggest, nitrogen is found in fluids mainly as NH_4 [27] and is therefore less susceptible to degassing. In short—so the argument runs—subduction-zone plate tectonics is the critical factor in defining the Earth's atmosphere and hence the planet's habitability by virtue of the essential role of nitrogen in structuring and sustaining life [13, 44].

Whatever the correct explanation, the issue has many ramifications. Studies of the conditions that led to the emergence of life need to specify the composition of the atmosphere (or water body) in which the synthesis took place. Much early such work assumed that strongly reducing conditions prevailed: that is, where the atmosphere contained molecules saturated with hydrogen atoms which were thus able to prevent or reverse oxidation in other molecules. In the influential Urey-Miller experiment, for example, a mixture of methane (CH_4), ammonia and water was exposed to UV radiation and electrical discharges. Miller and Urey [28] argued that the Earth and the other minor planets started out, like the giant planets, with reducing atmospheres but, unlike the giants, with their lower temperatures and higher gravitational fields, they became oxidising through the loss of hydrogen. The current view is that the Earth's atmosphere was in fact rich in oxidants such as CO_2 and N_2 from the outset and therefore not conducive to the production of organic compounds as proposed by Miller and Urey (see Chap. 7).

Nitrogen is found in other planetary atmospheres but, apart from that of Titan, in trivial amounts. On Mars the atmospheric nitrogen level is 2.7 % (95.3 % for CO_2) but atmospheric pressure as a whole is 0.6 % of the Earth's. On the other hand the rover *Curiosity* found evidence of fixed nitrogen in the form of nitrates in windblown deposits and mudstone into which it had drilled. The nitrogen is considered indigenous, the product of thermal shock resulting from impacts or lightning triggered by volcanic activity, and indicative of a nitrogen cycle at some stage in the planet's history [39].

To judge from this and other missions, as well as analysis of martian meteorites, the atmosphere has changed over time through differential loss to space [41] and (as with the methane pulses) combination with surface materials, but the planet's CO_2 and water appear to have reached their present levels ~4 Gyr ago [12] and this may also apply to its nitrogen. *Curiosity* also detected pulses of methane (CH_4) ten times the background level (0.001 % the Earth's level) which were open to a variety of interpretations besides the tempting biological one including the interaction between water and rocks containing olivine [32].

Secondary atmospheres are not always primary ones which have been amended by dirty impacts or gradual, differential leakage to space. Some are wholly novel, the gift of collision or volcanism on a surface that has been swept clean. The Earth's atmosphere may be an example of this, as the original tenuous atmosphere of hydrogen and helium released from its undifferentiated interior would have been blown away by the solar wind. In due course the Earth's new atmosphere came to be shielded from the solar wind, now greatly weakened as the Sun grew out of its juvenile T tauri phase, as well as from ultraviolet solar radiation, by the generation of a magnetic field, for which a metallic, liquid, rotating outer core was a prerequisite.

Comets and asteroids have long been mooted as a major source of the Earth's water. The deuterium/hydrogen (D/H) ratio, introduced in Chap. 2, appears to have resolved the matter after a false start. It will be recalled that four comets, including P/Halley, yielded a D/H ratio very different from that of the Earth's oceans [17] whereas the ratio for 103P/Hartley 2 resembled it. In 2015 analysis of comet 67P/Churyumov-Gerasimenko [3] by an instrument aboard the ESA's Rosetta spacecraft (Fig. 8.6) gave a value far higher than the ocean mean.

Now D/H ratios for asteroids such as carbon-rich chondrites and meteorites originating in the asteroid Vesta (Fig. 5.3) agree closely with that for oceanic water, adding weight to the notion that the early Earth was already rich in water. One plausible explanation is that water adsorbed on grains in the accretion disk could have delivered to the Earth 1–3 times the volume of its present oceans. But it remains to seen what proportion would have survived accretion [9] and, more crucially as regards this discussion, whether the oceanic D/H value has remained sufficiently unchanged for all the comparisons to have any validity. Indeed, one set of calculations suggests that the oceanic D/H increased by a factor of between 2–9 mainly because of gas interaction with a hydrogen-rich atmosphere [14].

Subsurface oceans further illustrate the complex outcomes of planetary differentiation. The ocean inside Saturn's moon Enceladus formerly thought to be confined to its southern polar area, has been shown by data from the Cassini mission to be of global extent [42] and its resistance to freezing explained by tidal effects due to Saturn. The icy plumes that had revealed Enceladus as a geologically active body do more than that: their highly alkaline composition prompts the suspicion that ocean water circulating through mantle rocks rich in iron and magnesium interacts with them ('serpentinisation') (see Chap. 7) and also liberates hydrogen, which promotes the formation of amino acids and also nourishes possible methane-producing bacteria [15].

Volcanism is of course a familiar outcome of differentiation, its progress depending on such matters as the development of an inhibiting crust, the concentration of radiogenic elements, and planet size [38]. Plate tectonics on Earth prolongs differentiation by forming crust and promoting mantle mixing. Mass redistribution arising from volcanism may lead to surface deformation and, in extreme circumstances, to orbital change. In a recent analysis of Martian geodynamics [4] it was suggested that the mass of the large volcanic edifice of Tharsis brought about a change in the planet's rotational axis by 20–25° leading to an obliquity of 45° which in turn displaced Mars' climatic zonation and drainage system.

Planetary magnetism also depends on distinctive zonation, as in models which invoke a liquid outer core. For the Earth, with a field that has operated for at least the last 3.5 Gyr, the dynamo has been powered by the cooling and freezing of the inner core, and a combination of experimental and computational analyses suggests that the inner core is less than 500–600 Myr old [8]. The progress of differentiation can only be traced thus: piecemeal and indirectly.

References

1. Ahrens TJ (1993) Impact erosion of terrestrial planetary atmospheres. Annu Rev Earth Planet Sci 21:525-555
2. Alfvén H, Arrhenius G (1976) Evolution of the solar system. NASA, Washington DC
3. Altwegg K et al (2015) 67P/Churyumov-Gerasimenko, a Jupiter family comet with a high D/H ratio. Science 347. DOI: 10.1126/science.1261952
4. Bouley S et al (2016) Late Tharsis formation and implications for early Mars. Nature 531:344-347
5. Cawood PA, KrönerA, Pisarevsky S (2006) Precambrian plate tectonics. Criteri and evidence. GSA Today 16:4-11
6. Citron RI, Zhong SJ (2012) Constraints on the formation of the Martian dichotomy from remnant crustal magnetism. Phys Earth Planet Inter 212:55-63
7. Condie KC, Kröner A (2008) When did plate tectonics begin? Evidence from the geologic record. In Condie KC, Pease V (eds) When did plate tectonics begin on Planet Earth? Geol Soc Am Spec Pap 440: 281-295
8. Davies C, Pozzo M, Gubbins D, Alfé (2015) Constraints from material properties on the dynamics and evolution of Earth's core. Nature Geosci 8: 678-685
9. Drake MJ (2010) Origin of water in the terrestrial planets. Meteoritics Planet Sci 490: 519-527

10. Dwyer CA, Nimmo F (2013) Chemical and Hf/W isotopic consequences of lossy accretion. 44 Lunar Planet Sci Conf, Abs 1773
11. Encrenaz T, Bibring J-P, Blanc M (1990) The solar system (2nd ed). Springer, Berlin
12. Fortes AD (2013) in Vita-Finzi and Fortes , Planetary Geology (2nd ed). Dunedin, Edinburgh
13. Galloway JN et al (2004) Nitrogen cycles: past, present and future. Biogeochem 70:153-226
14. Genda H, Ikoma M (2008) Origin of the ocean on the Earth: early evolution of water D/H in a hydrogen-rich atmosphere. Icarus 194: 42-52
15. Glein CR, Baross JA, Waite JH (2015) The pH of Enceladus' ocean. Geochim Cosmochim Acta 162: 202-219
16. Goldschmidt V (1937) The principles of distribution of chemical elements in minerals and rocks. J Chem Soc for 1937: 655-673
17. Hartogh P et al (2011) Ocean-like water in the Jupiter-family comet 103P/Hartley 2. Nature 478: 218-220
18. Hazen R et al (2008) Mineral evolution. Amer Miner 93: 1693-1720
19. Hitchcock DR, Lovelock JE (1967) Life detection by atmospheric analysis. Icarus 7:149-159
20. Jackson CRM et al (2013) Noble gas transport into the mantle facilitated by high solubility in amphibole. Nature Geosci 6:562-565
21. Jacobsen SB (2005) The Hf-W isotopic system and the origin of Earth and Moon. Annu Rev Earth Planet Sci 33:531-570
22. Kidman G, MacLean JS, Maxwell D (2014) Evidence of large scale tectonic processes on the Tharsis Rise, Mars. The Compass 86:2
23. Kleine T et al (2005) Hf-W chronometry of lunar metals and the age and early differentiation of the moon. Science 310:1671-1674
24. Laughlin GP, Bodenheimer P, Adams F-C (2004) The core accretion model predicts few Jovian-mass planets orbiting red dwarfs. Astrophys J 612:L73-L76
25. Lee D-C, Halliday AN (1997) Core formation on Mars and differentiated asteroids. Nature 388:854-857
26. Lewis JS (1997) Physics and chemistry of the solar system (rev ed). Academic, San Diego
27. Mikhail S, Sverjensky D (2014) Nitrogen speciation in upper mantle fluids and the origin of Earth's nitrogen-rich atmosphere. Nature Geosci 7:816-819
28. Miller SL, Urey HC (1959) Organic compound synthesis on the primitive Earth. Science 130:245-251
29. Militzer B (2013) Equation of state calculations of hydrogen-helium mixtures in solar and extrasolar giant planets. Phys Rev B 87:014202
30. Montmerle T et al (2006) Solar system formation and early evolution: the first 100 million years. Earth Moon Plan 98: 39-94
31. Müller RD et al (2008) Age-spreading rates and spreading symmetry of the world's ocean crust. Geochem Geophys Geosyst 9, DOI: 10.1029/2007GC001743
32. Oze C, Sharma M (2005) Have olivine, will gas: serpentinization and the abiogenic production of methane on Mars. Geophys Res Lett 32:L10203
33. Price NJ (2001) Major impacts and plate tectonics. Routledge, London
34. Rubie DC et al (2015) Accretion and differentiation of the terrestrial planets with implications for the compositions of early-formed Solar System bodies and accretion of water. Icarus 248: 89-108
35. Ruiz J 2014 The early heat loss evolution of Mars and their implications for internal and environmental history. Sci Rep 4: 4338
36. Schaefer L, Fegley B Jr (2010) Volatile element chemistry during metamorphism of ordinary chondritic material. Icarus 205:483-496
37. Schenk WB, Mckinnon WB (1989) Fault offsets and lateral crustal movement on Europa: evidence for a mobile ice shell. Icarus 79:75-100
38. Spohn T (1991) Mantle differentiation and thermal evolution of Mars, Mercury, and Venus. Icarus 90: 222-236

39. Stern JC et al (2015) Evidence for indigenous nitrogen in sedimentary and aeolian deposits from the Curiosity rover investigations at Gale crater, Mars. Proc Nat Acad Sci USA 112:4245-4250
40. Swindle TD (1997) Differentiation. In Shirley JH, Fairbridge RW eds, Encyclopedia of Planetary Sciences. Kluwer, Dordrecht
41. Taylor FW (2010) Planetary atmospheres. Oxford Univ Press, Oxford
42. Thomas PC (2016) Enceladus's measured physical libration requires a global subsurface ocean. Icarus 264:37-47
43. Tuff J et al (2013) Volcanism on Mars controlled by early oxidation of the upper mantle. Nature 498:342-345
44. Valley JW et al (2014) Hadean age for a post-magma-ocean zircon confirmed by atom-probe to mography. Nature Geosci 7:219-223
45. Yin Q et al (2002) A short timescale for terrestrial planet formation from Hf-W chronometry of meteorites. Nature 418:949-952
46. Zuber MT (2001) The crust and mantle of Mars. Nature 412:220-227

Chapter 6
Operation

Abstract The behaviour of the larger bodies, comets and meteorites making up our solar system owes much to the regularities imposed by Newtonian gravity but is influenced by random impacts and the complications introduced by migration, chaotic behaviour and the consequences of orbital resonance.

Aristotle's model of the cosmos was perhaps the most extravagant in antiquity as it invoked 55 concentric spheres on which were mounted the stars and planets (Aristotle Metaphysics). Earlier proponents of the model envisaged 26 (Eudoxus) and 33 (Callippus) crystalline spheres; Aristotle's were composed of aether and were individually powered by a god. Concentric motion in the same sense could not account for changes in the brightness and apparent size of some bodies or for retrograde motion—when a body appeared to change orbital direction—and later schemes, notably that of Ptolemy in the 2nd century AD, made use of epicycles—smaller circles added to the main orbit—to satisfy observation.

With newtonian mechanics, and in particular its inverse square law, the motion of individual bodies could be explained and, as shown later, unknown bodies identified by their distorting effect on measured orbits. All accepted the newtonian gravitational model of the solar system and its explanation of the elliptical orbits around the Sun that are followed by most orbiting planetary bodies and that are encapsulated in Johannes Kepler's first law of planetary motion (Fig. 4.3). It took Richard Feynman to point out, in a Caltech lecture the transcript of which was lost between 1964 and 1993, that Newton's explanation for the pervasive elliptical orbits could not be understood without knowledge of some 'arcane properties of conic sections'. Feynman did not know those properties and 'cooked up' a proof of his own using plane geometry [7].

In that lecture, Feynman showed that Kepler's second law—that the area swept out by a line from the Sun to the orbit would be the same during equal intervals of time—did not depend on the inverse square law. All it required was that the Sun should exert a pull, gravitational or something else, on the planet. Feynman claimed that his proof was purely geometric. Elsewhere he argued that, although the gravitational force might have immense explanatory power, we lacked a

© Springer International Publishing Switzerland 2016

C. Vita-Finzi, *A History of the Solar System*, DOI 10.1007/978-3-319-33850-7_6

mechanism for it [1]. But, like many other proofs, it invoked a constant of gravitation, usually represented by G. This remained an unexplained source of action at a distance which even Newton resented.

Moreover certain observations of planetary behaviour, notably changes in Mercury's orbit, were incompletely explained by classical gravity and they awaited Albert Einstein's general theory of relativity [5] in which gravity was transformed from a force to a distortion of spacetime. Alexander Pope's couplet

> Nature and Nature's laws lay hid in night:
>
> God said, Let Newton be! and all was light

thus turns out to be a little too upbeat. But for nonrelativistic technology Newtonian physics still reigns supreme, witness the triumphs of space exploration.

What had been learnt in the succeeding century about the role of gravity in the equilibrium and motions (or as we would now say the statics and dynamics) of the fluids and solids making up this and other solar systems was distilled into his *Mécanique céleste*, or Celestial Mechanics [12], by Laplace, whom we met in Chap. 1 as one of the exponents of a nebular model for the early solar system. Despite his evident admiration for Newton (whose name is the first word in the five volume *Mécanique*) we can already detect the Gallic suspicion of Anglo-Saxon empiricism in his rejection of all but indispensable data.

The data he did allow, however, revealed to Laplace that the system of the world was not rigidly predetermined but oscillated around a mean state. Newton too had detected that the interaction between the planets would perturb their orderly behaviour and he invoked God when necessary to restore the equilibrium and save the solar system from destruction. For this he was accused by Leibniz of disrespect for the clockmaker [3]. Compare Laplace's celebrated reply to Napoleon's enquiry over God's absence from the *Mécanique* reported in Chap. 1.

Only with general relativity do we have something approaching an explanation for the gravitational constant G: as a distortion of space-time, so that the planets orbiting the Sun are not being pulled by the Sun but are following the curved space-time deformation caused by the Sun. Later still came models based on gravitons or on gravitational waves but broadly speaking the newtonian structure for our solar system still prevails for many practical purposes, though sometimes after einsteinian tweaking. For example, taking special as well as general relativity into account, the clocks on GPS satellites need to run at 10.229999999543 MHz rather than at 10.23 MHz to allow for the distorting effect of gravity and of movement on time measurement [18].

The Stability of our Solar System

It is sometimes argued that the larger members of our solar system had acquired their present characteristics and gross arrangement perhaps as much as 4–4.5 Gyr ago [6] and that, although material subject to plasma physics continues to evolve, as regards the planets and larger planetesimals and comets subsequent events amount to tinkering with the details.

Fig. 6.1 Babylonian astronomical diaries record observations of moon and planets from the 7th century BC. Diaries for 164–163 BC appear to record the passage of Halley's comet 22–28 September 164 BC. Image by J Lendering, British Museum, courtesy of Creative Commons

The question of solar system stability is not straightforward. Take Henri Poincaré's essay on the subject. In 1887 King Oscar of Sweden announced a mathematical competition on this very subject, and the victor was the great mathematician. But after the winning essay had been published in the prestigious *Acta Mathematica* for 1899 its author spotted a serious mistake in his calculations, recalled all copies of the journal and pulped them, and issued a fresh version the following year. The original paper had demonstrated—by confining discussion to two unequal planets orbiting the Sun—that the system would be eternally stable. The revised version proved that the system was inherently unstable [20]. Not that the error was wholly sterile: in making the correction Poincaré supposedly was confronted by an early example of chaos in dynamical systems [3].

Present-day assessments of the solar system's working owe something to chaos theory as well as to relativity, but even without a conceptual revolution the certainties of Newtonian orbits could only survive scrutiny while planetary observation was confined to crude instruments and manual computation and until the aberration of light, the nutation ('nodding') of the Earth's axis and the proper motion of the stars were taken into account and planetary masses were adequately estimated. Newton's theory of universal gravitation implied that the planets had to perturb one another's motions, but planetary tables showed little sign of this effect into the 1780s [23] as the effect is small.

In fact the assumption of a well ordered dynamical system paid off when predicting future states or reconstructing past conditions, witness the routine forecasts of lunar eclipses by Chinese astronomers at least since the 2nd millennium BC and of solar eclipses, when the Moon's shadow subtends an angle of only 1/20° of arc, by Greek astronomers of the 2nd century BC. The dating of Babylonian texts (Fig. 6.1) by reference to their astronomical content [19] also appeals to a stable system, as does the discovery of at least two new bodies from anomalies in the orbits of one or more known planets.

The most celebrated example led astronomers to the planet Neptune on the basis of discrepancies between the calculated and observed orbits of Uranus. Urbain le Verrier, who in 1846 was the first to make the successful prediction (within 1°), found that the gravitational effect of the known planets could not account for the precession of the orbit of Uranus. Le Verrier and John Couch Adams, who also calculated the location of Neptune with good accuracy, used Titius-Bode's law as a starting point for their calculations even though the orbital distance it gives Neptune is now known to be 30 % too great.

The search for additional planets was not satisfied by the discovery of Neptune, partly because it did not wholly account for the anomalies in Uranus' orbit and partly because a further body seemed to be indicated by telescopic observations and theoretical calculations based on Laplace resonance. Percival Lowell drove the hunt of what he called Planet X but Pluto, discovered in 1930, proved a disappointing candidate as its mass proved too small to affect the giant planets. In any case data obtained by the Voyager 2 spacecraft in 1989 were to show that the mass of Neptune was 0.5 % smaller than assumed in conjectures about Planet X. In 1992 the discrepancies in Uranus' orbit were dismissed as erroneous [15].

Improved technology and modelling continued to add to the suburban population of the solar system to the tune of 35,000 trans-Neptunian icy bodies [10]. Moreover, the influence of a body with a mass close to that of Uranus—a new Planet X—appears to be indicated in the outer reaches of the solar system by six objects with elliptical orbits at an angle to the ecliptic plane (Fig. 6.2) [2]. The orbital period of this object is put at 10,000–20,000 yr and at its maximum reach it would extend the newtonian limits of our solar system to 1200 AU, well within the estimated limits of 50,000–20,000 AU estimated for the outer Oort cloud [17].

The present tally of solar system bodies (April 2016) is 8 planets, 5 dwarf planets, 178 moons, 3319 comets and 667,452 asteroids (NASA, 4 April 2016): as Alexander Humboldt [9] observed in *Cosmos*, planets (of which 11 were recognised in his day) with their moons numerically make up a very small proportion of the bodies in the solar system. The ratio is vastly greater today now we have the myriad Kuiper and Oort icy bodies. The observation is not trivial, as it bears on a feature used in the definition of planet adopted by the International Astronomical Union in 2006: a body which, besides being massive enough for its own gravity to make it 'round', has cleared its neighbourhood of smaller objects. For, whichever model is favoured for initial planetary gestation (see Chap. 3), planetary growth and survival fed on piracy and cannibalism.

In fact external gains and losses as well as internal rearrangement have featured in planetary studies from antiquity, as comets and meteorites were routinely observed even if their role was commonly coloured by superstition. Some perturbations were suspected of having a cumulative effect and were termed secular as opposed to cyclic. For instance, Edmond Halley thought that the observed reduction in Jupiter's orbit, contrasting with the expansion of Saturn's—something that had been noted by Kepler in 1625—might culminate in the entire loss of Saturn from the system and the destruction of the inner planets by Jupiter in its fall towards the Sun [3].

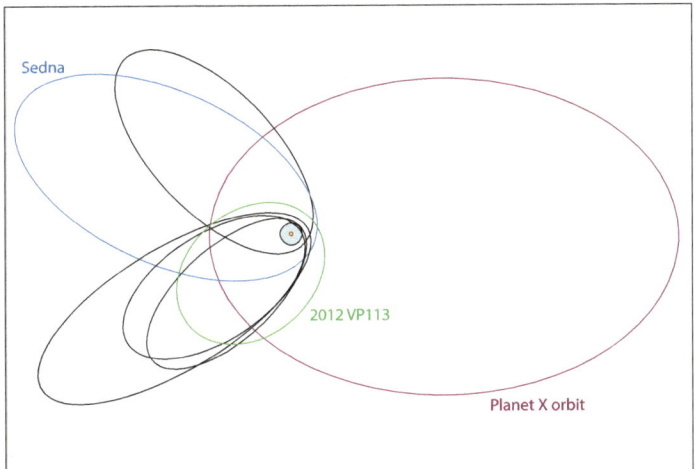

Sedna

2012 VP₁₁₃

Planet X orbit

Fig. 6.2 The dwarf planet Sedna, found in 2003, was then the most distant known object in the solar perihelion (the closest point to the sun) at 76 AU, beyond the Kuiper belt and far outside the influence of Neptune's gravity. The implication was that a massive body well beyond Neptune had pulled Sedna into its orbit, the prime suspect being the dwarf planet 2012 VP$_{113}$ orbiting the Sun far beyond Pluto, discovered in 2012. The new Planet X, announced in 2016, has a highly elongated orbit of 15,000 yr and may represent a planetary core (besides Jupiter, Saturn, Uranus and Neptune) which was ejected to its distant eccentric orbit by propinquity to Jupiter or Saturn [2]. Courtesy of Nature Publishing Group

On the other hand, according to Laplace the Jupiter-Saturn 'great inequality' between the orbits of Jupiter and Saturn, though apparently cumulative, was in fact cyclic with a period of 929 years [23]. Other, more unambiguously secular effects included a suspected increase in the Moon's orbital velocity, which (according to Kepler's second law) implied a reduction in its orbital radius and which seemingly resulted from progressive deceleration in the Earth's rotation produced by tidal friction. A decrease in the obliquity of the ecliptic—the angle between the Earth's axis of rotation and its orbital plane—since the third century BC also prompted discussion of secular versus cyclic changes.

A way of producing significant changes in the solar system while respecting its general stability is by collisions [16]. None better to illustrate this ruse than Newton, who in a reported conversation with the husband of his niece John Conduitt revealed that he thought the inner planets of the solar system would be devastated by a comet crashing into the Sun and causing it to expand dramatically and that the 1572 and 1604 supernovas were the result of the same process happening in other star systems. Three decades earlier he had told the mathematician David Gregory which the satellites of Jupiter were proto-Earths that were 'held in reserve' by God to repopulate the solar system after the cometary cataclysm [11].

Impact can shatter embryo planets; thus a simulated collision between two early giant planets gave rise to the terrestrial planets, the Moon and numerous smaller asteroids and comets [24].

If mild enough, however, collision can contribute to aggregation. Orbital chaos would seem to ensure collisions between, say the Earth and the other inner planets, and advances in computing have recently allowed direct analysis of the question. One such study relied on extended numerical simulation of 2501 orbits on the JADE supercomputer with present-day conditions as initial conditions. It showed that, although planetary perturbations could lead to distortion of the orbits of Mercury, Venus, Mars and Earth without close encounters or collisions, in 1 % of the cases the orbit of Mercury might be distorted enough to permit collision with Venus or the Sun which, in the space of 3.34 Gyr, in turn can completely destabilise the inner solar system and in some circumstances lead to the ejection of Mars or collision between Mars or Venus and the Earth [13]. In short, collisions between planets and planetary ejections are possible (with the relatively high probability of 1 %) in the remaining 5 Gyr life of the solar system.

Orbits, Tides and Impacts

The operation of our solar system thus combines the outcome of a complex evolutionary history with interaction between the constituent bodies and the geometry and violence of successive impacts. Hence the wide range of orbital parameters in the solar system and the problems that face any attempt to identify from them the process that gave rise to the planetary body in the first place. Consider the influence on rates of rotation of tidal forces and even (in the case of Earth and Mars) climatic effects as manifested in fluctuations of ice cover at sensitive latitudes.

Of course as orbits decay and may be subject to rounding the ideal would be some indication of the forces that have been at work; but even then there is scope for rival interpretations of the data. Take our Moon: as noted in Chap. 3, all are agreed that its synchronous rotation—that is its rotation with a period equal to that of its orbit—results from tidal friction of a prograde satellite orbit on a primary planet, but, whereas G. H. Darwin [4] and many of his contemporaries argued that it had been ejected by tidal forces, the current consensus is that impact expelled the material that came to form the Moon.

There is also disagreement over the tempo of the tidal effect. This acts both on the Moon and on the Earth (Fig. 6.3), and both on the oceans and on the solid Earth. A reduction in rotation rate, which is also manifested as an increase in the length of the day (l.o.d.) , is shown by growth bands in Devonian (400 Myr old) corals, which indicate 419 days a year and thus a day length of 21 h [21] while ~400 solar days a year, and thus a day length of 22 h, were derived from tidal rhythmites dating from ~620 Myr. [22] Combined with the periodicity indicated by banded iron deposits dating from 2450 Myr, the rhythmite data indicate a lunar recession rate of 1.24 cm/yr for much of the period 2450–620 Myr.

The current lunar recession rate as measured by laser ranging for 1970–2012 was 3.82 cm/yr. Why the rate should be irregular or show a net acceleration is not

Fig. 6.3 View of Moon and Earth taken by the Clementine spacecraft as it came over the northern lunar pole in March 1994. (The angular separation between lunar horizon and Earth has been reduced). The large crater at the bottom of the image is Plaskett (180° W longitude, 82° N latitude). Courtesy of NASA

clear: recession is of course a consequence of a reduction in the Earth's angular momentum which needs to be compensated by an increase in the Moon's and hence in its orbital radius (see Chap. 8).

By virtue of their orbital location Venus and Mercury presumably also experience tidal braking though solar in origin, and this may account for their slow spin rates (respectively 243 and 59 days) compared with those of the other planets. It is also likely that Mercury cannot attain synchronous rotation (which would be 88 days) because its rotation period is blocked, being in 3/2 resonance with its orbital period 'spin-orbit resonance'). Venus' rotation, in contrast, is faster than the synchronous rate of 224 days.

Why Venus' and Uranus' rotation should be retrograde is a question which, interesting in its own right, brings out the ramifications of what at first appeared a simple matter of frictional retardation, and illustrates the mix of history and accident that governs the solar system. Four explanations have been offered: the loss of a satellite, which was destroyed by an impact or which escaped perhaps to become Mercury; simple evolution influenced by solar tidal torque; the operation of retrograde rotation from the outset; and evolution from rapid, direct rotation to the present slow retrograde rotation [8]. So far no simple answer is forthcoming.

What of obliquity, a major control a planet's receipt of solar radiation as regards latitude and seasonality? Venus and Uranus have substantial obliquity, respectively 180 and 98°, but the thermal effect is complicated by eccentricity and by resonance between other orbital components. Moreover obliquity may change periodically, as in the Milankovitch cyclic model of climatic history for Earth and Mars, and numerical studies suggest that all the terrestrial planets may have been subject to large chaotic variations of obliquity at some time. Mars is still in this

Fig. 6.4 Ganymede
Jupiter's moon Ganymede, the largest moon in the solar system, displays faults and ridges as well as evidence of bombardment by asteroids, meteoroids and comets. Mapping based on images collected by NASA's Voyager and GalileoGalilei, spacecraft. Courtesy of NASA

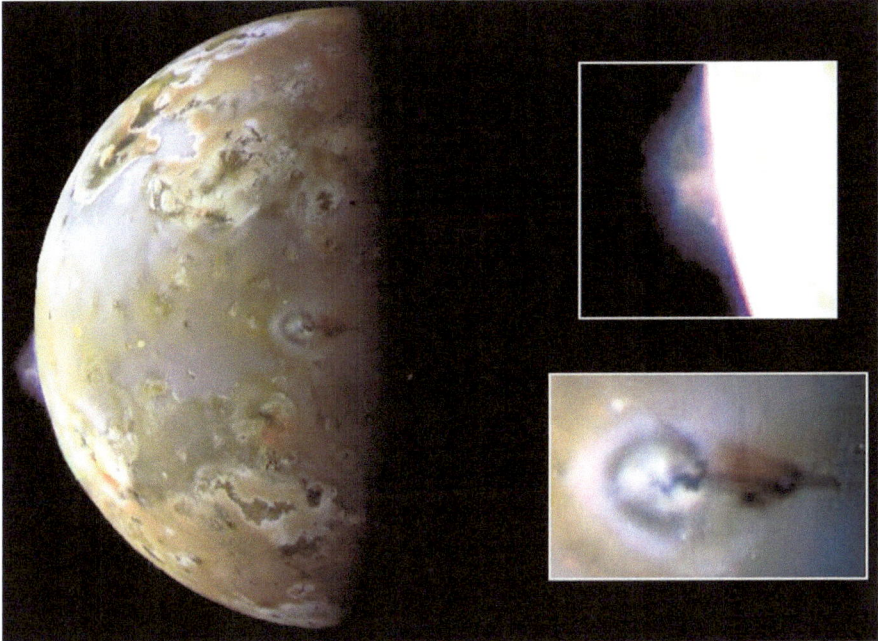

Fig. 6.5 Two eruptions are visible on Jupiter's volcanic moon Io: over its limb, a *bluish plume* rises about 140 km above the surface of the volcanic caldera Pillan Patera. Near the night/day shadow line, the ring shaped *Prometheus plume* is rises about 75 km. Image recorded on June 28, 1997 from a distance of about 600,000 km, from the Galileo spacecraft. Courtesy of NASA/JPL

situation over a range of 60°; Earth's obliquity was stabilised (though not eliminated) by its acquisition of a moon [14].

Tidal interaction deforms the two bodies at issue and can result in heating. Initially Io would have had a minute eccentricity and, being in synchronous rotation, any tides raised on it by Jupiter would not have affected its interior. But the tides would have displaced Io until its orbit fell into 4:2:1 resonance with Europa and Ganymede, (Fig. 6.4), whereupon it acquired an elliptical orbit along which it experienced tides of varying strengths which greatly flexed its interior and rendered it volcanically hyperactive (Fig. 6.5) in less than 500 Myr [25].

References

1. Alfvén H, Arrhenius G (1976) Evolution of the solar system. NASA, Washington DCDL.
2. Batygin K, Brown ME (2016) Evidence for a distant giant planet in the solar system. Astron J 151:22, doi: 10.3847/0004-6256/151/2/22
3. Brush SG (1996) Nebulous earth. Cambridge Univ Press, Cambridge.
4. Darwin GH (1898) The tides and kindred phenomena in the solar system. Houghton, Boston.
5. Einstein A (1915) Erklärung der Perihelbewegung des Merkur aus der allgemeinen Relativitätstheorie. Preuss Akad Wiss Sitzungs 1915, part 2, 831–839.

6. Feynman RP, Leighton RB, Sands ML (1963) The Feynman lectures on physics I. Addison-Wesley, Redwood City CA.
7. Goodstein DL, Goodstein JR (1996) Feynman's lost lecture: the motion of planets around te Suin. Norton, New York NY.
8. Henrard J (1991) Celestial mechanics, in McNally D (ed), Reports on astronomy, Trans Int Astr Un 21A, 15–27.
9. Humboldt A de (1850) Cosmos. Italian trans of 1845. Turati, Milan.
10. Jewitt DC, Luu JX (1995) The solar system beyond Neptune. Astr J 109: 1867–1935.
11. Keynes Ms (2004) Account of a conversation between Newton and Conduitt (1725). Keynes MS 130.11, King's College Cambridge.
12. Laplace PS de (1798) Traité de mécanique céleste. Duprat, Paris.
13. Laskar J, Gastineau M (2009) Existence of collisional trajectories of Mercury, Mars and Venus with the Earth. Nature 459: 817-819.
14. Laskar J, Roboutel P (1993) The chaotic obliquity of the planets. Nature 361: 608-612S.
15. Levenson T (2015) The hunt for Vulcan: how Einstein destroyed a planet and deciphered the universe. Random House, NY.
16. Melita MD, Woolfson MM (1998) A numerical algorithm for dissipative Keplerian discs. MNRAS 299: 60-72.
17. Morbidelli A (2008) Origin and dynamical evolution of comets and their reservoirs. arXiv:astro-ph/0512256v1.
18. Selleri F ed (1998) Open questions in relativistic physics. Apeiron, Montreal.
19. Stephenson FR, Steele JM (2006) Astronomical dating of Babylonian texts describing the total solar eclipse of s.e.175. J Hist Astron 37: 55-69.
20. Villani C (2012) Théorème vivant. Grasset, Paris.
21. Wells JW (1963) Coral growth and geochronometry. Nature 197: 948-950.
22. Wiliams GE (2000) Geological constraints on the Precambrian history of Earth's rotation and the Moon's orbit. Rev Geophys 38: 37-60.
23. Wilson C (1985) The great inequality of Jupiter and Saturn: from Kepler to Laplace. Arch Hist Exact Sci 33: 15-290.
24. Woolfson M (2013) A postulated planetary collision, the terrestrial planets, the Moon and smaller solar-system bodies. Earth Moon Plan 111, DOI:10.1007/s11038-013-9420=8.
25. Yoder C (1979) How tidal heating in Io drives the Galilean orbital resonance locks. Nature 279: 767-770.

Chapter 7
Life

Abstract There is growing evidence for precursors of biomolecules in many parts of the solar system and, to judge from telescopic imaging from the ground and from space as well as from sampling of meteorites and comets, elsewhere in our galaxy and beyond its confines. Their age can be estimated on Earth and more controversially on Mars, while conditions that appear propitious for life can be identified and sometimes dated on other solar system bodies and on a number of exoplanets. The interaction between living and abiotic processes can now be traced in broad terms over much of the Earth's 4.6 Gyr lifetime, to the benefit of both geology and evolutionary biology.

The search for life elsewhere in the solar system, mooted since antiquity and sustained from time to time by scraps of controversial evidence, was energised by remotely sensed evidence for episodes of surface water on ancient Mars and for subsurface seas on Europa, Enceladus and perhaps also Ganymede, Titan, Callisto and Triton. It now benefits from the discovery and analysis of extrasolar planets which are considered to be orbiting within their parent stars' habitable zone, from a growing list of terrestrial organisms—extremophiles—capable of surviving seemingly hostile environments and from evidence for biomolecular precursors on comets, extragalactic star systems and interstellar space. The search also benefits from conceptual advances which have broken free from narrow views of the nurturing environment and of the crucial processes, as well as from the recognition that 'life is the evolutionary outcome of a process and not a single fortuitous event' [40].

But perhaps as important has been the change in attitude towards the concept of panspermia, a term once applied to the delivery of life or at any rate its precursors to the Earth, but now used more loosely for the notion that life is commonplace in the universe. In either guise it is viewed as a feeble solution to the difficulties confronting biochemists seeking to create life in the laboratory and, despite the support of two giants of science, the physicist Francis Crick and the astronomer Fred Hoyle, and Kelvin with his 'seed-bearing meteoric stones' before them, it seemed merely to outsource the key problem without solving it.

C. Vita-Finzi, *A History of the Solar System*, DOI 10.1007/978-3-319-33850-7_7

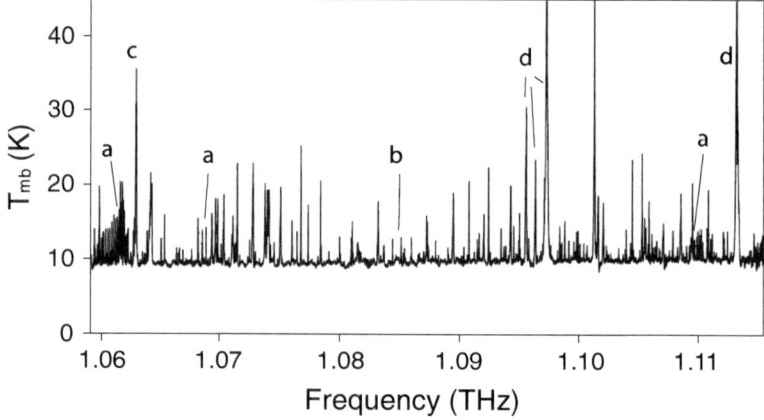

Fig. 7.1 The HIFI spectrum of the Orion Nebula: among molecules identified in this spectrum are methanol (**a**), carbon monoxide (**b**), hydrogen cyanide (**c**) and water (**d**). Image credit @ Cassini Imaging Team, SSL, JPL, ESA and NASA

The likelihood that some amino acids would survive large cometary impacts [59], experimental evidence for the survival of bacteria in impacting meteorites [14], and most graphically the apparent survival on the Moon of *Streptococcus mitis* from Surveyor 3 thirty years after its journey there [53], have shown that interplanetary travel by microrganisms, once derided, is at any rate feasible. Not that the search for mechanisms to account for the emergence of life on Earth has in any way abated, fuelled as it is by the recognition of hitherto unsuspected biochemical pathways [64], while the possibility of extraterrestrial contributions to the mix has strengthened at least one current model (Fig. 7.1).

The acceptance of extraterrestrial life is not simply linked to a particular religious doctrine. Democritus (c. 460–370 BC) suggested that the universe contains other planets supporting life. Giordano Bruno, whose brutal fate was mentioned in the introduction, was one of a succession of philosophers for whom the restriction of life to Earth seemed incongruous in the face of an apparent multiplicity of celestial earths. William Herschel maintained that the Sun was a solid planet and was convinced that a cooler surface beneath its hot visible exterior was inhabited. Giovanni Schiaparelli's 1877 report of Martian canali, mistranslated as canals, perpetuated the myth of Martian civilisation well into the twentieth century mainly through the advocacy of Lowell [44]. Today we need to be reminded that the ability to synthesise some of the building blocks of life does not guarantee 'that life itself is a preordained inevitability' [19].

Life and its Context

Definitions of life are numerous but many of them are concerned more with language than biology (see LIFE/LEBEN) [74], and it has been argued that controversy will persist until the formulation of a theory of the nature of living systems [18].

Fig. 7.2 False-colour
UV image of double haze
above Saturn's moon Titan,
an opaque atmosphere
composed mainly of nitrogen
and a detached haze probably
containing ethane, CO_2 and
other organic molecules.
Image credit @ Cassini
Imaging Team, SSL, JPL,
ESA and NASA

Operational definitions have their own limitations [7], however, and the issue of definition can be sidestepped by focusing on the interaction between the putative life forms and their response (or indifference) to particular environmental stimuli: for example, a lack of oxygen or exposure to ultraviolet light. It is a conservative, cautious approach which risks missing novel or alien lifeforms but can be supplemented if new data lead to enlightenment.

The need remains for flexibility over the constraints: thus liquid water is often cited as a prerequisite for all forms of life, but ethane and methane, both of them liquids that are stable on the $-179\,°C$ surface of Saturn's moon Titan (Fig. 7.2), could serve as biosolvents and underpin an alternative, self-sustaining, replicating, organic chemistry [9, 46]. Other assumed limitations may also prove to be illusory: the puzzling success of a green sulfur bacterial species in a black smoker plume at deep-sea hydrothermal vents on the East Pacific Rise that are devoid of sunlight is explained by the exploitation of geothermal radiation which includes wavelengths at the margin of the visible and infrared spectrum (viz ~390–700 nm) but which is successfully absorbed by the bacteria's 'photosynthetic' pigments [5].

Similarly the definition of what is lifeless needs constant readjustment, as was made clear when the Precambrian, a stretch of Earth history which had long been viewed as sterile, was found to pullulate with microbial life [63]. That the Hadean Eon (4.56–3.8 Gyr ago) was abiotic may thus prove a premature assessment: quite apart from any relevant progress in molecular palaeontology which might reveal latent lifeforms there is the tantalising fact that, even though it is an informal unit not based on stratigraphy, the Hadean is better known from lunar than from terrestrial evidence; indeed its onset is defined in some accounts by the impact between the early Earth (or Tellus) and the hypothetical body Theia that gave rise to the Moon, and is preceded by the Chaotian Eon, which embraces the events that gave rise to the solar system. In short, the Hadean connotes hellish conditions which were to be found elsewhere.

Now, whereas the environment—terrestrial, aquatic and atmospheric—was formerly given the dominant role in accounts of Earth history, with living species having to adapt to changing circumstances, evolution is now seen to have serious geological consequences to the point where the two realms, animate and inanimate, may be said to coevolve [31, 37]. In the 550 Myr of the Hadean eon there may have been to hand a mere 420 rock-forming or accessory (i.e. subordinate) mineral species which were widely distributed or significant in volume or both. Compare the ~4800 post-Hadean mineral species that subsequently arose in response to vol-canism, tectonics, impacts and the many other processes that enriched the Earth's surface and subsurface mineralogy [32]. The contrast owes much to the processes mediated by life, such as the limestones and other products of biomineralization.

The level of intimacy between the two strands is of course open to discussion. Consider the interplay between microbial and chemical activity that gave rise to stromatolites, those mounds or columns of laminated sediment built up in shallow water by cyanobacteria in the course of at least the last 2.4 Gyr [12]. The pro-cess as manifested, for example, in Australia's Shark Bay involves photosynthesis which, by withdrawing carbon dioxide from the water, promotes the precipitation of calcium carbonate which in turn traps grains of sediment and helps to build up successive layers [61]. And changes in the environment led to variations in stroma-tolite type, well displayed in sections exposed along Siberian rivers [36].

But this example is a poor one as it illustrates aeons of complacent domesticity rather than of ambitious coevolution. A better (and much grander) example used to illustrate geobiology, though sometimes projected as a geological product of evolution rather than as a two-way process, is oxygenic photosynthesis. A revised version of photosynthesis, it fed oxygen to the atmosphere and oceans, so that the Earth's environmental history can thus be divided into three major episodes: the Archaean (>2.5 Gyr), with little or no free oxygen, the Proterozoic, with some oxygen in the atmosphere and the upper oceans, and the Phanerozoic (<542 Myr), with the oceans entirely oxygenated [37].

One of its consequences was to make sunlight accessible as a free energy source thus liberating life from dependence on environments such as hydrothermal vents where localised, strong reduction-oxidation (redox) gradients can power the conversion of carbon dioxide into organic matter [26, 36]. Submarine hydrother-mal systems were seriously considered as potential sites for the origin of life, or at any rate its survival [1], following the discovery in 1979 of a flourishing fauna at the vents on the Galápagos Rift in the East Pacific at a depth of 2600 m and at a temperature of 17 °C [20]. Besides their chemical and thermal attributes the vents added to the range of extremophiles a new category free from the constraints of dependence on solar light by relying on bacterial chemosynthesis—the conver-sion of chemicals into usable energy. And it is the source, not necessarily the sole source, of Earth's oxygen.

Fig. 7.3 Tubeworms of the Pogonophora family at a large sulphur and sulphide emitting vent on Axial Seamount, the most active submarine volcano in the NE Pacific. It rises to a depth of 1400 m below sea level and erupted in 1998, 2011 and 2015. Vent temperatures were 35 °C, around 30 °C hotter than the surrounding waters. Other vents nourish bacterial mats, smaller tube worms, sea snails, sea spiders and limpets. Image courtesy of NOAA/PMEL

Some 200 currently active submarine rift systems have been identified in the 40 years since the East Pacific sites were discovered (Fig. 7.3), and many more await confirmation—a situation reminiscent of the exoplanet story, when the first recognition in 1992 was followed in a mere 24 years by over 1600 confirmed sightings with about 4700 Kepler candidates [55]. Hydrothermal vents appear ephemeral relative to geological time scales, as they depend primarily on active interplate rifting (Fig. 7.4) but the vent habitat appears to have existed throughout the Phanerozoic (the last 542 Myr) and perhaps since the Archaean (4–2.5 Gyr ago) and thus offered time for adaptation, while species analysis supports the notion that migration was along mid-ocean ridges rather than by the shortest oceanic route and is therefore closely linked to the development of ocean spreading centres. For example, ridge connections across the Pacific during the early to mid-Cenozoic (65–30 Myr) may largely account for the strong similarity at genus and family level between vents in the NE and W Pacific despite their current separation [71].

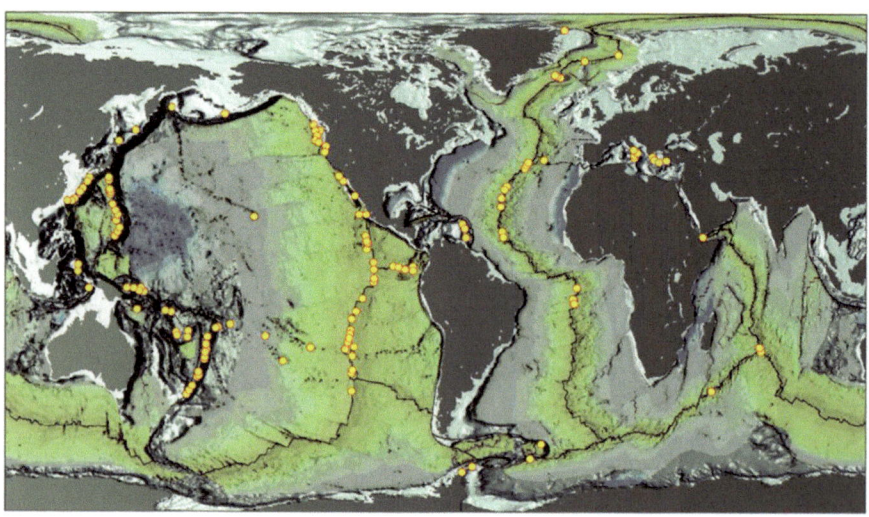

Fig. 7.4 Distribution of active hydrothermal vent fields in 2010 after [6] plotted on map of seafloor age [48]. Both the activity of individual fields and the pattern of rifts are subject to continual change

True, repeated visits to individual black smokers reveal major structural changes over days to decades [5], as one might expect given that some rifts spread at rates as high as 20 cm/year so that the locus of maximum activity is likely to have shifted and some smokers doubtless shut down permanently. Larval dispersal is influenced by oceanic circulation as well as larval development time so that, whereas basin-to-basin dispersal may occur only once in tens of thousands of years, the strong Kuroshio current may link vent fields that are 1200 km apart [52]. In fact, shallow, hypersaline inshore settings are nowadays, at any rate, apparently more uncommon than hydrothermal vents, witness the scarcity of live stromatolite colonies.

Nevertheless the loss of potential habitable sites, specialised or banale, is clearly a secondary concern if life on a planet has been drastically reduced. Extinctions were long accepted with great reluctance, a product of the uniformitarian tradition which saw Earth history as predominantly measured and unmelodramatic [10]. Yet they had long served to subdivide the fossil record because they marked relatively sudden changes in the faunal content of the rocks. As with panspermia there came about a change in perception which was only partly driven by fresh evidence. A key event was undoubtedly the publication of a paper headed by Walter and Luis Alvarez which pointed to the environmental consequences of an asteroid impact [2]. The change was given a helpful shove by the demonstration by Shoemaker [68] that Barringer Crater in Arizona was the product of a meteorite much of which had vaporised on impact, and even then many years passed before the craters on the Moon were generally accepted as creatures of impact rather than volcanicity.

In some discussions, mass extinction is defined as the loss of 20 % or more of all species globally [34] but this criterion is by no means straightforward notably as regards the circumstances preceding and following the event. Even the five major mass extinctions that are widely recognised—end-Ordovician, late-Devonian, end-Permian, end-Triassic and end-Cretaceous, with the end-Permian (a 60 Myr interval about 252 Myr ago) the most severe, [15] appear arbitrary when expressed by loss or gain in genus diversity, and a concern with degrees of partial loss shows up these matters as parochial when viewed from the viewpoint of the solar system as a whole: total planetary sterilization is required as the basis for global stratigraphy. True, that then requires life to be restarted, but as the next section suggests, this obstacle is not insurmountable.

Geologists working on fossil-free chunks of Earth history have successfully demonstrated that there are many abiotic routes to setting up a framework into which life and its migration can eventually be fitted but, at present, events marked by the emergence and extinction of species still far outnumber events dated by radioisotopes in the early stratigraphic record [66]. The problem of identifying marker beds or chronohorizons is all the greater for interplanetary correlation for the simple reason that, as things stand, life has hitherto been identified only on Earth.

Radiometric ages provide robust chronological links between the Earth, Moon and Mars for the first 100 Myr of solar system history: one of ~4.3 Gyr on terrestrial minerals, 4.36 Gyr on an anorthosite on the Moon, 4.4 on Martian meteorite NWA 7433, and 4.21 ± 0.35 Gyr on a mudstone in situ on Mars. For later times stratigraphic correlation between the inner planets based mainly on crater density shows little chronological agreement even though at least some of the major cratering events were probably shared. The mismatch in one instance doubtless arises at least in part because some workers assume that any craters older than 0.5 Myr on Venus were completely obliterated by an episode of volcanic resurfacing for which the evidence is in fact open to challenge [72].

Straightforward comparison by crater counting (Fig. 7.5) is in any case hampered by differences in gravity and atmospheric density between different bodies [29]. Using signals arising from solar fluctuations faces a similar problem, for, as noted in Chap. 4, different bodies respond differently to flares or changes in net insolation.

Hence the proposed recourse to an extrasolar signal: a preliminary attempt used as marker horizon the fallout from a supernova explosion which had left its mark on Earth in the shape of a ^{60}Fe signal in a ferromanganese crust at a depth of 4830 m in the Equatorial Pacific [35]. Type II carbon deflagration supernovae are the main source of this isotope in our galaxy, which can travel over distances of 50 pc or ~160 light years. The event was put at 2.8 Myr (2.1 Myr on the basis of a revised half life ($t_{1/2}$) for ^{60}Fe of 2.62 Myr, [22]) and reports of ^{60}Fe elsewhere in an eastern Pacific core from magnetite crystals in magnetotactic bacteria in ocean-floor material dating from 2.2 Myr [11] suggest the isotope may serve as a terrestrial chronomarker. The cosmic ray spectrum points to a nearby supernova about 2 Myr ago [13], and ^{60}Fe is now reported from all the major oceans in deposits dating from 1.5–3.2 Myr and 6.5–8.7 Myr ago [73] so that, rather like volcanic ash beds, ^{60}Fe horizons can be used to correlate widely separated sequences on Earth and on other solar system bodies, freeing crater chronologies from dating

Fig. 7.5 Crater count based on elevation data returned by the LOLA instrument on the Lunar Reconnaissance Orbiter. Colour coding shows deviation from global 'mean sea level' of 1737.5 km of craters >20 km confirming long standing conclusion that any primeval bombardment of the nearside basins had been obliterated by huge impacts and lava flows. Of the craters in the census, *green* mean global elevation; *bluer* below, *yellower* above. Courtesy of NASA/GSFC/ LOLA/Brown/SVS

duties. But more significant for the present discussion is the recovery of ^{60}Fe in lunar samples at levels which are inconsistent with production by galactic or solar cosmic rays [22] and which, once dated independently (not an easy matter in the shallow, disturbed lunar regolith), may prove to originate in the same supernova activity and thus provide a valuable interplanetary marker horizon.

More supernova signals will doubtless be found, as the rate of supernova explosion in our local galactic neighbourhood within some 100 pc from Earth is estimated to be one per 2–4 Myr [73]. Known supernovae whose physical trace has yet to be found include the AD 1054 event that produced the Crab Nebula and Kepler's supernova of 10 October 1604. It would be useful to know whether

[60]Fe can be supplemented in the hunt by other indicators especially if they can be accessed remotely and thus on bodies besides the Earth.

At the very least one can only hope that the lunar link will be substantiated and complemented by comparable finds on Mars, and on other solar system bodies, at more than one point in time. The search for markers produced by supernovae will benefit from improved laboratory techniques and from the deployment of isotopes such as plutonium 244 (^{244}Pu) applicable to marine sediments which offer better time resolution than a ferromanganese crust [60]. In similar vein, a sharp increase in cosmogenic ^{14}C in tree rings dated to AD 775 has been ascribed to a galactic gamma-ray burst [58] or a solar energetic particle event, although the lack of supporting astronomical evidence is troubling [69].

That there is scope for such markers is clear from a consideration of the galactic history of our solar system. The Sun has revolved around the Galaxy about 20 times since its formation, and there may have been consequences of biological interest in the flux of ionising radiation, in the accretion of gas and dust during passage through interstellar clouds, and in impact cratering rate [21] all of which could have left a record on the Moon and on other receptive bodies. The last would have been modulated by changes in the gravitational environment, notably when passing above and below the orbital plane and through spiral arms, which would perturb the orbits of Oort cloud comets and hence the cratering rate in the inner solar system.

This in turn may serve as a link between extinctions and planetary chronology [42] and reinforce the craving among some geoscientists (including astrobiologists) for cycles by adding ~26–37 Myr, the half-period of the Sun's oscillation about the Galactic plane, [4] to the list of periodicities identified in extinction rates [34].

Planetary sterilization, the optimum punctuation for planetary biostratigraphy, remains undocumented. But it is sometimes taken for granted in discussions of the battering by meteorites suffered by the Earth (and presumably its neighbours) about 4 Gyr ago. Wholehearted extinction is also invoked in the context of solar superflares, stellar events releasing up to 10^7 more energy than the largest recorded solar flare (10^{32}erg) which have been recorded on main-sequence stars [62]. Superflares are considered unlikely to occur on Earth because they appear to result from magnetic reconnection between a star and a Jupiter-sized planet with a close orbit and, less convincingly, because global aurorae have not been reported over the last 500 years and 'appropriate extinctions' have not occurred over the last ~1–2 Gyr [62].

Origins

The search for the earliest lifeforms on Earth began, naturally enough, with a hunt for the earliest fossils, a category soon broadened to include bacterial products such as algal mats and carbonate constructs. The next step was to seek isotopic signatures of living processes [54], a procedure allied to James Lovelock's dismissal of extant life on Mars because the planet's atmosphere composition did not betray the state of disequilibrium to be expected from interaction with life [33]. The microfossil record goes back about 3.5 Gyr, but carbon isotopic evidence for biological activity, which derives from the fractionation resulting from enzimatic carbon fixation, takes it much further [8].

The enterprise now embraces much of the solar system, which is seen to provide a wide range of potential cradles and changing environments, and the new science of astrobiology (or exobiology) demonstrates with its name that space in its entirety deserves to be explored and that panspermia, with its emphasis on the transmission of seeds, and abiogenesis, a term which focuses on the emergence of life from inanimate matter, are concepts too grand to be rooted solely in our earthly experience.

The search has been pursued, though at a modest level, throughout scientific history, with the ancient Greeks as often asking pertinent questions. In the 6th century BC, for example, Anaximander proposed wet soil drying out as a potential cradle, a scheme that prefigures today's clay-mineral models [16]. The concept of spontaneous creation was supposedly demolished by Louis Pasteur in 1864 when he showed that sterile solutions remained uncontaminated if sheltered from microorganisms in the air. But it was revived by Leduc [41] as a necessary part of evolution, and in 1953, when Stanley Miller showed that, in the presence of water vapour, passing a spark through a mixture of gases which were then thought by some to have been present on the early Earth—methane (CH_4), ammonia (NH_3), and hydrogen (H_2)—yielded a number of the amino acids found in proteins [17]. The same procedure on a non-reducing atmosphere dominated by N_2 and CO_2 (with oxidation inhibited by the addition of ferrous iron) also yielded significant amounts of amino acids [17].

The Miller experiment is sometimes held up simply as putting into effect the 'primordial soup' model formulated by Oparin and Haldane [28, 57]. But it is much more than that: by showing that biological molecules could be produced experimentally and that abiological preconditions could be manipulated in the light of new information. 'The spontaneous generation of life...is extremely improbable (wrote Lovelock [43] ... (with) the implication that wherever life exists its biochemical form will be strongly determined by the initiating event ... (which) ... could vary with the planetary environment at the time of initiation'. The fact that prebiotic synthesis has been successful in a wide range of chemical environments does not signify that the task is an easy one, but it suggests that simple prebiotic compounds—and perhaps also more complex molecules—are common in the cosmos [3].

The next step, to an accepted definition of life, generally remains problematic, but not if we adopt a gradational definition which ranges from unfeeling bacteria to self-conscious humanity, especially if we consider its environmental variety and the resilience and adaptability of microorganisms. And if we embrace viruses in the living world—even if Lwoff [47] claimed they were neither organisms nor molecules but simply viruses—then the step is imperceptible.

But even if its culmination is a matter of debate—and owes much to contingency [45]—the need remains to start the process. Current models can be crudely grouped into four overlapping categories: biochemical (along the lines of Miller's 1953 experiments and its numerous progeny), biophysical (with an emphasis on membrane evolution) [70], energetic (as with proton gradients [38]) and informational (primarily focusing on RNA and DNA: [24]). The overlap is especially evident in the role given to hydrothermal systems, currently much in favour through the pioneering advocacy of Szostak et al. [70], where the precipitation of an inorganic membrane comprising iron sulphides and oxyhydroxides is a key step at alkaline submarine vents, and abiotic chemistry leads to chemoautrophic prokaryotes (i.e. single-celled organisms deriving their free energy from oxidation of components of the environment such as nitrate and ferric iron rather than from solar energy) and thence to nucleated cells. As argued by Wächtershäuser [75], instead of self-assembly of such structures as RNA and proteins in a soup already containing suitable components, there might take place metabolism in a high-temperature, high-pressure setting rich in iron sulphide.

Volcanic vents have since been joined as candidate settings by non-volcanic hydrothermal systems [39, 49]. The interest in this variant of the chemoautrophic approach resides in its greater permissiveness: as we saw earlier, seafloor spreading and its hot venting is common but localised and discontinuous; the more low key process of serpentinisation [65] where the mineral olivine is converted by exothermic reaction with water to the snakeskin-like serpentine, and in so doing generates hydrogen fuel for emergent life as well as heat which also leads to moderate temperatures (~100–130 °C rather than 350 °C) but without being confined to seafloor spreading zones let alone a single hydrothermal vent.

But the permissiveness is still circumscribed by geological history. It may be no coincidence that the Earth ~4 Gyr ago boasted rapidly spreading mid-ocean ridges and active volcanic arcs, with a hotter mantle (and thus lavas) thanks to higher levels of radioactivity than today as well perhaps as residual heat from planetary accretion [56]. Whether the circumstances went so far as to promote chemoautrophic activity the terrestrial record is a useful clue to possible life haunts elsewhere [23].

The continuing controversy over Martian meteorite ALH84001, found in Antarctica in 1984, where it fell about 11,000 year ago after its ejection from Mars ~17 Myr ago, demonstrates the limitations of morphological and geochemical evidence for past life on another solar system body. The meteorite displayed five potential life indicators: bacteria-like shapes, though much smaller than most bacteria on Earth, magnetite crystals aligned as if in terrestrial magnetotactic bacteria; PAHs; carbonate globules indicating moderate temperatures and water; and minerals within the globules which are commonly, though not uniquely, associated with life [50]. But a further assessment of ALH84001 and two other meteorites with oxygen isotopic composition and trapped gases characteristic of Mars—Shergotty and Nakhla—showed that, although the evidence was 'compelling', it did not satisfy all the criteria that have come to be adopted for identifying past life in a geological sample [27].

Such criteria are generally rooted in a conservative definition of life, that is one circumscribed by Earthly experience, where the record is itself contaminated by the existence of life. Hence the value of an extraterrestrial context for understanding the origins of life on earth as well as elsewhere. 'We cannot distinguish the contingent from the necessary' [67]. It seems sensible to go back one step in the inorganic-organic progression and focus on the transition between inorganic and organic molecules.

Evidence has accumulated over the last century for amino acids and other organic compounds in comets, including Wild 2, in meteorites, such as Murchison, and in stellar systems, such as HR 4796A in Centaurus. PAHs too, as we saw in Chap. 2, are widely represented in the Universe, including planetary nebulae [25] and closer to home in the upper atmosphere of Titan (Fig. 7.2). There would seem to be grounds for viewing life as potentially universal. Other things being equal, and that includes obliquity and planetary dimensions, the major discriminant could turn out to be the cosmic ray flux, which in turn hinges on the shielding provided by magnetic fields and atmospheres.

References

1. Abramov O, Mojzsis SJ (2009) Microbial habitability of the Hadean Earth during the late heavy bombardment. Nature 459:419-422
2. Alvarez LW, Alvarez WA, Asaro F, Michel HV (1980) Extraterrestrial cause for the Cretaceous-Tertiary extinction. Science 208: 1095-1108
3. Bada J (2013) New insights into prebiotic chemistry from Stanley Miller's spark discharge experiments. Chem Soc Rev 42: 2186-2196
4. Bahcall JN, Bahcall S (1985) The Sun's motion perpendicular to the galactic plane. Nature 316: 706-708
5. Beatty JT and 8 others (2005) An obligately photosynthetic bacterial anaerobe from a dep-sea hydrothermal vent. Proc Nat Acad Sci 102: 9306-9310
6. Beaulieu S, Joyce K, Soule A (2010) Global distribution of hydrothermal vent fields. At http://www.interridge.org/irvents/maps
7. Bedau MA, Cleland CE (2010) The nature of life. Cambridge Univ Press, Cambridge
8. Bell EA et al (2015) Potentially biogenic carbon preserved in a 4.1 billion-year-old zircon. Proc Nat Acad Sci 112:14518-14521
9. Benner SA (2010) Water: constraining biological chemistry and the origin of life. In: Lynden-Bell RM et al (eds) Water and life: the unique properties of H_2O. CRC Press, Boca Raton FL, 157-176
10. Benton M (2003) When life nearly died. London, Thames and Hudson
11. Bishop S, Egli R (2011) Discovery prospects for a supernova signature of biogenic origin. Icarus 212: 960-962
12. Blank CE (2004) Evolutionary timing of the origins of mesophilic sulphate reduction and oxygenic photosynthesis: a phylogenetic dating approach. Geobiology 2:1-20
13. Breitschwerdt D et al (2016) The locations of recent supernovae near the Sun from modelling Fe transport. Nature 532:73-76
14. Burchell MJ (2004), Mann JR, Bunch AW (2004) Survival of bacteria and spores under extreme shock pressures. Mon Not R Astron Soc 352: 1273-1278
15. Burgess SD, Bowring S, Shen S-Z (2014) High-precision timeline for Earth's most severe extinction. Proc Nat Acad Sci 111: 3316-3321

16. Cairns-Smith AG (1982) Genetic takeover and the mineral origins of life. Cambridge Univ Press, Cambridge
17. Cleaves HJ et al (2008) A reassessment of prebiotic organic synthesis in neutral planetary atmospheres. Preb Chem 38:105-115
18. Cleland CE and Chyba C (2002) Defining 'life'. Orig Life Evol Biosph 32: 387-393
19. Conway Morris S (1998) The crucible of creation. Oxford Univ Press, Oxford
20. Corliss JB et al (1979) Submarine thermal springs on the Galápagos Rift. Science 203:1073-1083
21. Crawford IA et al (2010) Lunar palaeoregolith deposits as recorders of the galactic environment of the solar system and implications for astrobiology. Earth Moon Plan 107:75-85
22. Fimiani L and 12 others (2014) Evidence for deposition of interstellar material on the lunar surface. Lunar Planet Sci Conf 45, abs 1778
23. Fishbaugh KEW et al (2007) Geology and habitability of terrestrial planets. New York, Springer
24. Fox-Keller E (2000) The century of the gene. Oxford, Oxford UP
25. García-Hernández DA and 7 other (2010) Formation of fullerenes in H-containing planetary nebulae. Astrophys J Lett 724: L39-L43
26. Gargaud M et al (2012) Young Sun, early Earth and the origins of life. Heidelberg, Springer
27. Gibson EK Jr and 5 others (1999) Evidence for ancient martian life. Fifth Mars Int Conf, Abs 6142
28. Haldane JBS (1929) The origin of life. Repr. in New Biol 16, 1954, 12
29. Hartmann WK (2005) Martian cratering 8: isochron refinement and the chronology of Mars. Icarus 174:294-320
30. Hazen RM (2009) The emergence of patterning in life's origin and evolution. Int J Dev Biol 53:683-692
31. Hazen RM (2013a) The story of Earth. New York, Penguin
32. Hazen RM (2013b) Paleomineralogy of the Hadean Eon: a preliminary species list. Am J Sci 3313:807-843
33. Hitchcock DR, Lovelock JE (1967) Life detection by atmospheric analysis. Icarua 7:147-159
34. Huggett RJ (2006) The natural history of the Earth. London, Routledge
35. Knie K et al (2004) ^{60}Fe anomaly in a deep-sea manganese crust and implications for a nearby supernova source. Phys Rev Lett 93, 171103
36. Knoll AH (2003) Life on a young planet. Princeton, Princeton Univ Press
37. Knoll AH (2009) The coevolution of life and environments. Rend Fis Acc Lincei 20: 301-306
38. Lane N, Martin W (2010) The energetics of genome complexity. Nature 467: 929-934
39. Lane N, Allen JF, Martin W (2010) How did LUCA make a living? Chemiosmosis in the origin of life. BioEssays 32: 271-280
40. Lazcano A (2010) Historical development of origins research. Cold Spring Harb Perspect Biol 2. DOI:10.1101/cshperspect.a002089
41. Leduc S (1911) The mechanism of life. Rebman, London
42. Leitch EM, Vasisht G (1998) Mass extinctions and the Sun's encounters with spiral arms. New Astron 3: 51-56
43. Loveloc3 k 1965 = 1969?Nature 207: 568-570
44. Lowell P (1908) Mars as the abode of life. McMillan, New York
45. Luisi PL (2015) Chemistry constraints on the origin of life. Isr J Chem 55:906-918
46. Lunine JI (2009) Saturn' Titan: a strict test for life's cosmic ubiquity. Proc Am Phil Soc 153:403-
47. Lwoff A (1957) The concept of virus. J Gen Microbiol 17: 239-253
48. Map of seafloor age after http://www.ngdc.noaa.gov/mgg/image/crustageposter.jpg. For earlier periods the planetary configuration would need adjustment
49. McDermott JM et al (2015) Pathways for abiotic organic synthesis at submarine hydrothermal fields. Proc Nat Acad Sci 112:7668-7672
50. McKay DS and 8 others (1996) Search for past life on Mars: possible relic biogenic activity in martian meteorite ALH84001. Science 273: 924-930

51. Miller SJ (1953) A production of amino acids under possible primitive earth conditions. Science 130: 528-529
52. Mitarai S et al (2016) Quantifying dispersal from hydrothermal vent fields in the western Pacific Ocean. Proc Nat Acad Sci USA 113:2976-2981
53. Mitchell FJ, Ellis WL (1971) Surveyor III: bacterium isolated from lunar retrieved TV camera . In Levinson AA (ed) Proc 2nd Lunar Conf 3:2721-2733. MIT Press, Cambridge
54. Mojzsis SJ (1996) Evidence for life on Earth before 3,800 million years ago. Nature 384:55-59
55. NASA Exoplanet Archive (18 April 2016) at Exoplanetarchive.ipc.caltech.edu
56. Nisbet EG (1985) The geological setting of the earliest life forms. J Mol Evol 21: 289-298
57. Oparin A (1924) The origin of life (Eng trans 1938: New York, Macmillan)
58. Pavlov AK and 6 others (2013) AD 775 pulse of cosmogenic radionuclides production as imprint of a Galactic gamma-ray burst. Mon Not Roy Astr Soc doi 10.1093/mnras/stt1468
59. Pierazzo E, Chyba CF (2010) Amino acid survival in large cometary impacts. Meteor Planet Sci 34:909-918
60. Raisbeck G et al (2007) A search for supernova produced ^{244}Pu in a marine sediment. Nucl Instr Meth Phys Res B 259: 673-676
61. Riding R (2011) The nature of stromatolites: 3500 million years of history and a century of research. In Reitner J et al (ed) Advances in stromatolite geobiology. Springer, Heidelberg, 29-74
62. Rubenstein EP, Schaefer BE (2000) Are superflares on solar analogues caused by extrasolar planets? Astrophys J 529:1031-1033
63. Rudwick M (2014) Earth's deep history. Chicago, Univ Chicago Press
64. Russell MJ, Martin W (2004) The rocky roots of the acetyl-CoA pathway. Trends Biochem Sci 29: 358-363
65. Russell MJ, Hall AJ, Martin W (2010) Serpentinization as a source of energy at the origin of life. Geobiology 8:355-371
66. Sadler PM (2006) Composite time lines: a means to leverage resolving power from radiosiotopic dates and biostratigraphy. Pap Paleont Soc 12: 145-170
67. Sagan C (1974) The origin of life in a cosmic context. Orig Life Evol Biosph 5: 459-505
68. Shoemaker EM (1959) Impact mechanics at Meteor Crater, Arizona. US Geol Surv, Open-File Rep
69. Stephenson FR (2014) Astronomical evidence relating to the observed ^{14}C increases in A.D. 774-5 and 993-4 as determined from tree rings. Adv Space Res 55:1537-1545
70. Szostak JW, Bartel DP, Luisi PL (2001) Synthesizing life. Nature 409: 387-390
71. Tunnicliffe V, Fowler CMR (1996) influence of sea-floor spreading on the global hydrothermal vent fauna. Nature 379:531-533
72. Vita-Finzi C, Howarth RJ, Tapper S, Robinson C (2004) Venusian craters and the origin of coronae. Lunar Planet Sci Conf 35, abs 1564
73. Wallner A et al (2016) Recent near-Earth supernovae probed by global deposition of interstellar radioactive ^{60}Fe. Nature 532:69-72
74. Wood MG et al (eds) (2014) Dictionary of untranslatables. Princeton, Princeton Univ Press
75. Wächtershäuser G (1990) Evolution of the first metabolic cycles. Proc Nat Acad Sci USA 87: 200-204

Chapter 8
The Evolving Solar System

Abstract Advances in the observation of the heavens from the ground and from satellites, novel computational techniques, and theoretical advances in a number of related (and some unrelated) fields permit changes in the solar system to be monitored in ever increasing detail. The greatest progress has been in analysing the dynamics of the Earth, the Sun's EUV radiation, and the inflow of comets and asteroids. The outcomes bear on the physical and chemical processes that modify the solar system.

The notion of a solar system running like clockwork is clearly misleading, unless it is a clock which is not entirely dependable or with grit in the works [7]. Isaac Newton (as mentioned in Chap. 1) saw the need for occasional divine intervention whenever the interaction between planets and comets upset the planetary orbits. In 1692 Edmond Halley likewise claimed Almighty Wisdom had made the Moon denser than the Earth to ensure it was not left behind in their passage through the viscosity of space [7]. The recognition of chance events in the assembly of the solar system [24], should have swept away the clockwork image, but it persists in those accounts where the solar system has a well defined date of birth. Yet, as preceding chapters have shown, assembly as well as adjustment have continued long after the Year Zero of the chronology derived for the Earth. This chapter shows how some of the resulting changes are being traced.

The enterprise is by no means novel. Besides solar and lunar eclipses, which have been recorded and forecast at least since Babylonian times (Fig. 6.1) [19], and which were clearly cyclical, a number of astronomical changes have long been recorded whose cyclic nature could not be recognised in the short term without a measure of audacity. Consider, for example, the precession of the equinoxes, studied by Hipparchus in the 2nd century BC, or the variable orbital motion of the Earth, which was apparently known to the manufacturer of the Antikythera mechanism, the 2nd century BC clockwork computer recovered in 1900–1 from a Greek shipwreck [5] (Fig. 8.1).

© Springer International Publishing Switzerland 2016   85
C. Vita-Finzi, *A History of the Solar System*, DOI 10.1007/978-3-319-33850-7_8

Fig. 8.1 The Antikythera mechanism, a device for calculating the solar and lunar calendar, eclipses, and the movement of other stars and planets; it includes epicyclic gearing and a slot-and-pin mechanism to allow for subtle variations in the Moon's motion across the sky. It was found in a Greek shipwreck it is dated to about 100–150BC. Some see the mechanism as epitomising a much earlier view of a mechanical universe than we traditionally accept (See Edmunds M et al 2014 The Antikythera mechanism—the book. Sitkas, Athens). Image (fragment A) about 18x15 cm, licensed under Creative Commons

The distinction between periodic, cumulative and irregular changes may hinge on the context and it could change with technology. Edmond Halley was perhaps the first to detect, using ancient eclipse data, that the Moon's orbital rate was accelerating, and many investigators have pursued the complex processes that govern what at first glance seems a straightforward phenomenon. It is not simply a matter of calibration: even after the introduction of Atomic Time in 1955, long-term trends could be determined with 'fair precision' only by relying on naked eye observations going back to about 700 BC, while acceleration of the Earth's rate of rotation were recognised in the mid-18th century, and variations in the daily rate or length of day (l.o.d) were demonstrated in 1939 [20].

The clocks that continue to reveal astronomical variations include water clocks, pendulum clocks, clockwork clocks and atomic clocks, themselves prone to error as regards accuracy, stability and, as regards the last groups, susceptibility to jitter, the name for variations in the frequency generated by the oscillator at the heart of the system. As clock instability owes much to temperature variations, satellite clocks may be cocooned in an onboard oven and their error may thereby be reduced by orders of magnitude. But the assessment will in turn depend on regular calibration and thus further crosschecking. The 200 or so atomic clocks that underpin International Atomic Time (TAI) are thought to deviate from precision by 1 s in 20 Myr, and they allow astronomical or Universal Time (UT1), originally based on Greenwich, to be calibrated and where necessary corrected by the use of a leap second. The last was added on 1 July 2015.

As elsewhere, such errors are clues to underlying processes as well as nuisances to be combated. The length of the l.o.d, for example, displays a progressive increase due to tidal braking which amounts to 2.3 ms per century. On it are superimposed fluctuations produced by the interaction between the Earth and the atmosphere, earthquakes, and the mass redistribution of oceans and ice bodies produced by seafloor spreading and glacial history.

The complexity of this issue [10], which bears on many other solar system problems, is illustrated by the 'enigma' of an apparent mismatch between the decrease in the Earth's rotation rate over the last three millennia and changes in global mean sea level reported for the 20th century [12]. The rotation rate derived from the timing of ancient eclipses, after allowing for ocean expansion due to warming and glacial unloading at high-latitudes, disagreed with the rate due to sea-level rise resulting from glacial melting combined with a 20th century component [2].

An important contribution to resolving the puzzle came from improvements in the historical eclipse record, and hence in estimates for slowing in rotation rate, thanks to a signal for angular momentum exchange between the mantle and the fluid outer core. The recent sea-level rise had also been overestimated, and its correction benefited the modelling of glacial isostatic adjustment in the shape of ice unloading and shifts in water load [11].

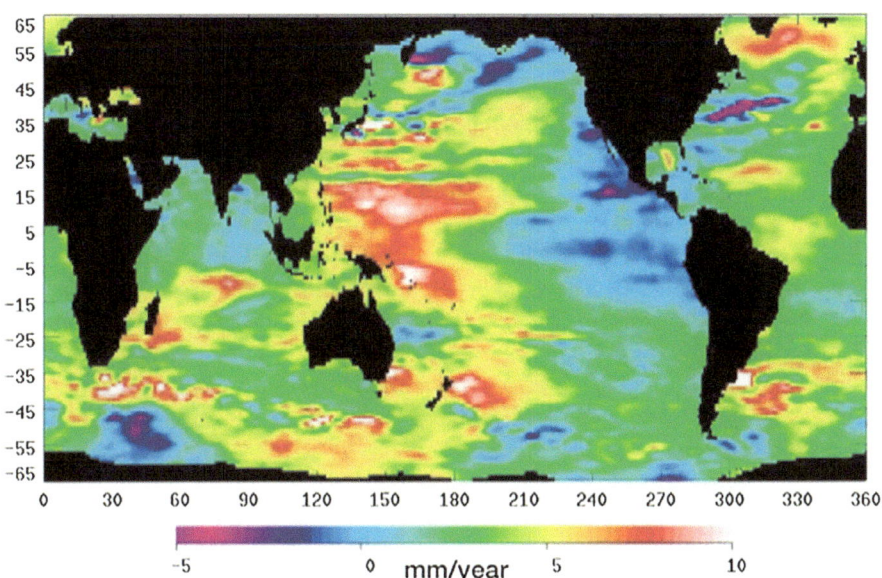

Fig. 8.2 Trend of sea level change 1993–2008 showing variability across the globe, with some areas showing no change and others a rise of up to 10 mm/year in response mainly to winds, currents and long-term changes in circulation including the Pacific Decadal Oscillation. Data from Jason-1 and Topedx-Poseidon satellites. Image PIA11002 courtesy of NASA/JPL

To refer baldly to sea-level rise is to ignore its regional variation as well as to its changing complexity over time. Thus despite continuing glacial melting the pace of sea-level rise has been slowed by 20 % thanks to the storage of 3.2 trillion tons of water on land in lakes and aquifers, an effect identified with the help of two satellites launched in 2002 for the Gravity Recovery and Climate Experiment (GRACE) [14]. What is more, modelling suggests that the rise will affect different parts of the globe differently, with submergence greatest in the Pacific and some polar areas enjoying emergence [18]. The outcome, plotted by satellite and thus no longer compromised by tide gauge disturbance (Fig. 8.2) shows that sea-level modelling still falls short of predicting local variations as its interpretation calls for evidence from geology, geophysics and isotopic chemistry as well as climatology and glaciology and cannot in any case recover long-term changes away from coasts.

The Sun

It is fitting that the most detailed and continuous piece of solar system monitoring should bear on the Sun, driven both by scientific curiosity and the need for sober assessment of the solar contribution to climatic changes, health and energy sources. The results bear both on current variations in solar output and on the Sun's inner workings.

Several dedicated spacecraft have been launched in recent decades to observe this or that component of solar behaviour, and they have revealed variability on many fronts (Fig. 8.3). Operating above the distorting atmosphere they have shown that the Sun's irradiance, which averages 1.36 kW/m^2, is not constant but increases by about 0.1 % at the peak of the 11-year sunspot cycle and varies by 6.9 % during the year thanks to the Earth's eccentric orbit. The Ulysses spacecraft (1990–2009), with an orbit which brought it three times over both solar poles, found that the solar wind during its third orbit was 25 % weaker than during the first orbit [9]. The total irradiance monitor (TIM) on the Solar and Heliospheric Observatory (SOHO, 1995–) contributes to the solar radiation and climate experiment (SORCE, 2003–) with daily and 6-hourly measures of TSI reported at 1 AU. Since 2010 EUV observations over 145 years by SOHO [15] are being followed by the Solar Dynamics Observatory (SDO, 2010–) at higher time resolution.

An improved grasp of the Sun's output of ultraviolet (UV) radiation should help us understand where in the Sun it is generated and to forecast its likely fluctuations. UV radiation, especially at the low end of its frequency range, is a major influence on both terrestrial and space weather, primarily through its action on the ionosphere and the upper atmosphere. Moreover UV-C (190–280 nm) and vacuum UV (<190 nm) wavelengths would probably be lethal to most organisms. But the radiation is blocked by the Earth's atmosphere and especially its CO_2, molecular oxygen and ozone (O_3) component. The lack of oxygen and therefore ozone on

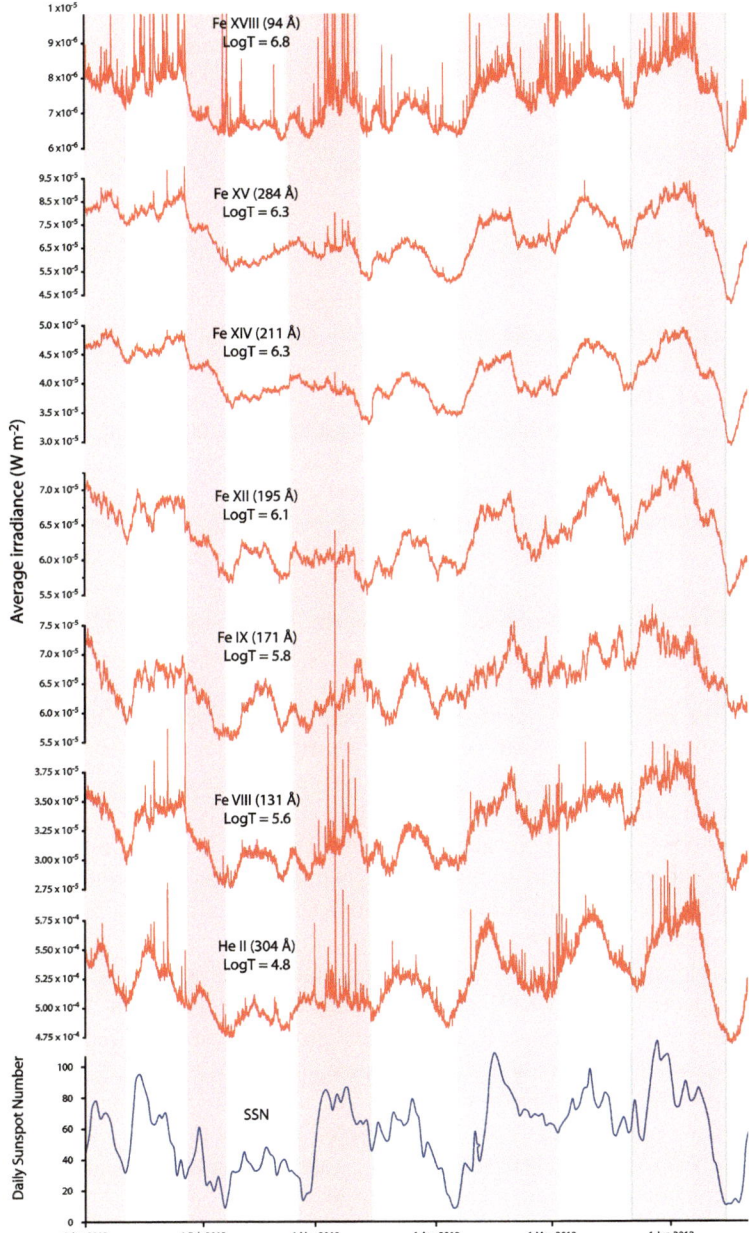

Fig. 8.3 Sample output of average irradiance for 7 coronal lines and net sunspot number (SSN) for the corresponding period. Coronal line 1 min averages plotted as 10 min moving averages. Data from EUV SDO variability experiment (EVE) at lasp.colorado.edu/ courtesy of NASA. SSN data for the visible disk according to the formula R = K (10g + s) where g is the number of sunspot groups and K is a scale factor intended to normalize the data to the scale introduced by J.R.Wolf in 1848, from http://ftp.ngdc.noaa.gov courtesy of NOAA

Earth in the Archaean, c 3.5 Gyr ago, as well as on early Mars, doubtless influenced early evolution both by directly promoting mutations and by favouring some organisms at the expense of others [15]. How far the dense Venusian atmosphere protected any former denizens by filtering out UV is still an open question as the age of its atmosphere is still not known let alone when it earned the label of greenhouse; similarly the issue of Mars's atmospheric history. Astronauts will of course need artificial protection and will doubtless rely on space weather forecasts for planning their missions.

As the Sun's UV flux fluctuates broadly in unison with total solar irradiance (TSI), it experiences the ~11-year sunspot (Schwabe) cycle, which is linked to differential rotation of the Sun, and a ~27 day oscillation, which is conventionally ascribed to rotation of the photosphere or visible surface of the Sun. But components of the EUV can vary over the cycle by as much as 160–300 % compared with 0.1 % for TSI as a whole, rendering sensible forecasts problematic.

There are substantial EUV satellite records for 1991–2002 obtained by the Solar Ultraviolet Spectral Monitor (SUSIM) and the Solar Stellar Irradiance Experiment (SOLSTICE), the latter replaced by a second SOLSTICE instrument since 2003. For earlier times reliance for UV data has to be placed on interpolation in TSI records, and a synthetic dataset ($E_{10.7}$) has been made available for collaborative research by Space Environment Technologies [25]. It represents the integrated solar EUV energy flux for the full solar spectrum at the top of the atmosphere since February 1947, and thus potentially for six solar cycles. This fine resource was used in an attempt to identify any periodicities in the EUV flux at 1.8–105 nm within a solar cycle. The shorter-term oscillations for 1993, an unremarkable year, have a periodicity of between 23 and 33 days depending on where the baseline is placed. Although this could merely reflect a rotational factor superimposed on the Schwabe cycle, solar rotation accounts for no more than 42.31 % of the variation [8] so that, even if we add a possible further 15 % due to annual changes in the Earth's location relative to the Sun in the course of a yearly orbit there remain 43 % to be accounted for.

A possible explanation is that a quasi-regular control resides in the solar interior, as suspected by Dicke [3] when he asked whether there was 'a chronometer hidden deep in the Sun', and that its effects are distorted by differential rotation and the vicissitudes of active regions. Neutrinos offered to provide some kind of window into the solar interior which, unlike most others, reflected present-day conditions: as noted in Chap. 4, it is accepted that photons may take thousands of years to ricochet to the surface so that, rather as with the stars above, much of what we see took place millennia or aeons ago.

As a glimpse of what may be eventually be accomplished, consider the Super Nova Early Warning System (SSNEWS): as neutrinos are released by a supernova in a few tens of seconds, whereas it may be hours or days before the first electromagnetic signal is detected, several neutrino observatories form part of the network to give astronomers advance warning of a supernova in our galaxy or in a neighbouring one [1].

The measured solar neutrino flux is famously skimpy and even that was for long suspected of poorly reflecting the established standard solar model (SSM) until the apparent deficit was explained by the failure of the existing gallium or chlorine devices to detect muon and tau neutrinos and only electron neutrinos. To counter the problem of scanty data it sometimes pays to combine time series. When such a strategy was adopted with data for two detectors (Homestake and GALLEX) there emerged a modulation with a sidereal frequency (i.e. rotation relative to the other stars) of 12.85 year^{-1}; and measurements from space by the ACRIM (Active Cavity Radiometer Irradiance Monitor) during the corresponding period gave an identical periodicity. This was boldly taken as evidence that irradiance is modulated by rotation of the solar core at that frequency [22].

How the two realms communicate is by no means clear especially in view of the complex circulation that characterises the Sun's interior. The Extreme Ultraviolet Variability Experiment (EVE) on the SDO satellite seemed to offer some scope for linking EUV fluctuations on the solar surface or photosphere to the rest of the Sun's atmosphere and the solar wind. The 6-month period 1 January–30 June 2012 was selected for preliminary study because it lies near the midpoint on the rising limb of Solar Cycle 24 rather than at an extreme position in activity. Figure 8.3 shows irradiance in watts m^{-2} as one minute averages plotted as moving averages for 7 coronal lines for 1 January–1 July 2012. The lines in the figure range in postulated formation temperature up to $10^{6.8}$ K, that is over 1 million degrees. The plots show broad synchroneity among the various lines combined with substantial variation in the timing of major peaks. The turning points in all the sequences are in fair agreement even if the details are not [26]. And there is also broad agreement with the sunspot record.

Fig. 8.4 The solar corona during a total solar eclipse. Parts of the chromosphere, which plays an important part in the model discussed here, can be glimpsed. Courtesy of NASA

A plausible mechanism is induction heating. Another instrument on the SDO, the Atmospheric Imaging Assembly (AIA), reveals numerous EUV cyclones at the solar surface which are presumably modulated by convection within the Sun. These induce currents in the chromosphere, which plays the part of ferro-magnetic cookware in which an eddy current is induced by an oscillating magnetic field in a subjacent coil and serves to convert magnetic to thermal energy. As Fig. 8.3 shows, their periodicities are transmitted throughout the corona (Fig. 8.4), and they link the solar interior to the solar wind and thus to much of the solar system and beyond.

Alien Invaders

As the solar wind and solar radiation in general demonstrate, the boundaries of the solar system are inevitably ill-defined, and, what is more, there is evidence of continuing invasion from wholly alien sources. This can be useful when an attempt is made to coordinate the geological history of the inner planets. The aim was to provide a framework against which one could compare the relative timing of major

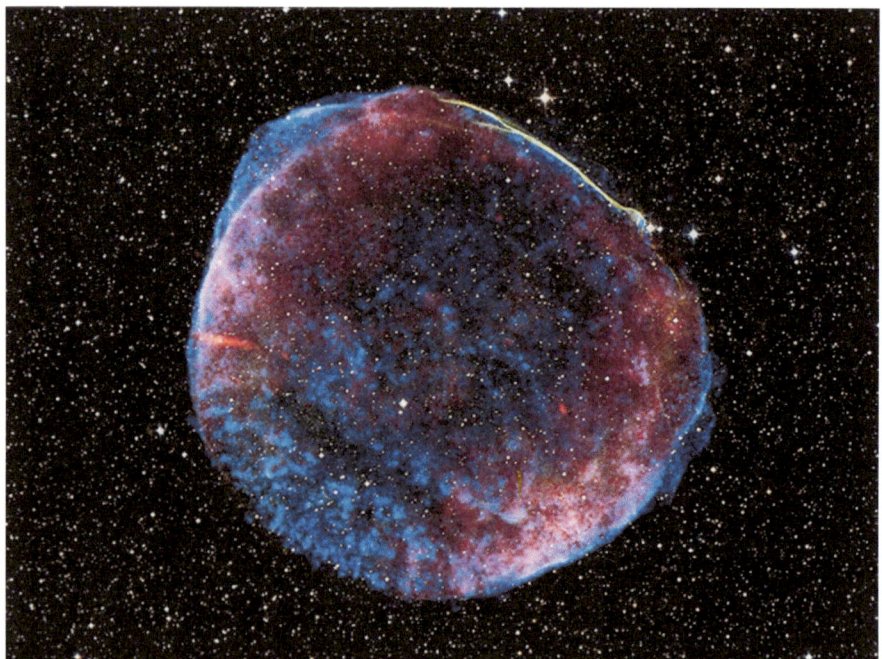

Fig. 8.5 The SN1006 supernova remnant, the remains of a white dwarf star which exploded 7200 light years before the light reached the Earth in AD 1006, when it was 10× as bright as Venus and recorded in Egypt, Switzerland, China and Japan. The energy spectrum detected on SN1006 is thought to provides strong evidence that surpenovae generate cosmic rays. Image courtesy of NASA, ESA and Zolt Levay (STSci)

events on different solar system bodies. Consider the assumption that all the inner planets and our Moon had shared in the Late Heavy Bombardment where, as T.H. Huxley observed in 1862, similarity of sequence or homotaxis is not necessarily evidence of simultaneity with synchrony. As noted in Chap. 3, even if the concept of a cataclysm survives it may refined by an improved chronology, and could eventually shed light on the nature or source of the impacts.

Numerical ages (probably radiometric) would be the ideal basis for time correlation across the solar system but they are too few even in the Earth-Moon system. Solar fluctuations are still too poorly understood to present identifiable events although the isotopic or chemical traces they might yield could eventually be traced on icy bodies. Crater counting (Fig. 7.5), calibrated with lunar numerical ages, has long been the favoured device, and it leads to dispute even in the case of Mars. Quite apart from the assumption that planets shared in the battering, the calculation is not straightforward especially as the craters can represent more than one episode and their number may be distorted by secondary features [6]. A source which offers some promise as a chronohorizon, as we saw with regard to mass extinctions (Chap. 7), is a supernova, as receipt of its signal would be close to simultaneous throughout the solar system.

Besides providing age markers these studies have shed light on a longstanding puzzle: the connexion cosmic rays and supernovae. NASA's Fermi Gamma-ray Space Telescope identified two supernova remnants, IC 443 and W44, where protons are being accelerated. A study by the Very Large Telescope of the European Southern Observatory reached the same conclusion with regard to SN 1006, a supernova remnant first recorded in the year 1006 (Fig. 8.5). Supernovae may not be responsible for all GCRs but we have here sound evidence that they provide some of them.

Fig. 8.6 Comet Philae lander of the ESA's Rosetta mission on the surface of Comet 67 Philae revealed much about P/Churyumov-Gerasimov, 12 November 2014. Philae has revealed much about the isotopic composition of the water on the comet and the presence of organic compounds. Image courtesy of ESA/Rosetta/Philae/CIVA

Thus supernovae have appeared in the narrative in three roles: as possible triggers for nebular collapse, as a potential means (via ^{60}Fe) of correlating events on different solar system bodies, and as a source of cosmic rays which, in their turn are seen as triggers for cloud condensation [23].

A related issue is how far the GCR flux to Earth is influenced by the heliosphere. When assessing the validity of dates based on cosmogenic isotopes such as ^{10}Be, it is seen as essential to allow for the distorting effect of the Earth's magnetic field and of the solar wind. Recent work shows that it is no longer sufficient to think in terms of the inner heliosphere [16]. For instance, the GCR flux may be doubled if the solar system passes into a dense unmagnetised interstellar cloud, a possible explanation for the enhanced radioisotope production indicated by the ^{10}Be record in Antarctic ice 35,000 and 60,000 years ago: [4] further evidence that, even if the solar system has not expanded, our interpretative boundaries necessarily have.

In any case it is increasingly clear that both major and minor shifts in the orbital location of the major bodies have occurred at all stages in solar system history through gravitational interaction and collision. And it is misleading to view the Oort Cloud and the Kuiper Belt as reservoirs of pristine spare parts or scraps: the numerous comets held in storage in the Kuiper Belt and the Oort cloud have been modified, sometimes substantially, by thermal, collisional, radiation and interstellar-medium processes, with collision processes much more effective in the relatively more crowded Kuiper belt. We may therefore still view the Oort comets as damaged relics of the time of formation of the outer planets, but the Kuiper

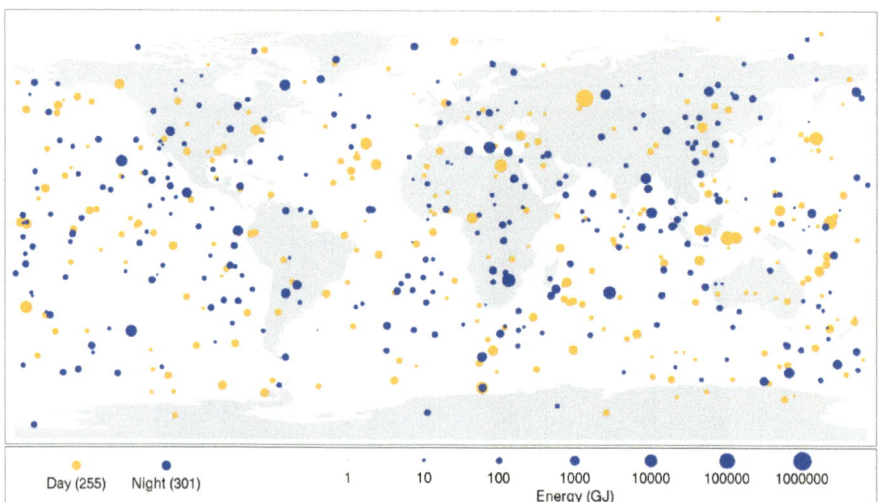

Fig. 8.7 Small asteroids which disintegrated in the Earth's atmosphere in 1999–2013. Note scale of energy liberated by the disintegration in gigajoules (GJ). 1 GJ = 278 kWh. Map courtesy of NASA

comets are now seen to be fragments chipped off larger Kuiper belt objects during the last 10–20 % of the lifetime of the solar system [21], and all comets are potentially reservoirs of information which bears on the solar system as a whole (Fig. 8.6).

Asteroids, likewise long viewed as little changed since the collapse of the solar nebula, are now known to have been scattered widely by migrating planets and to have been modified by space weathering, collision, and exposure to the solar wind [17]. Interplanetary dust particles (IDPs) are also subject to collision and solar radiation pressure, in particular the non-radial component known as Poynting-Robinson drag which causes fine dust to spiral into the Sun. Though the process may take 10^9 years to convey dust from the asteroid belt [13] the flux of meteors (Fig. 8.7) and the zodiacal light show that the dust is continually replenished.

In short, the magnetic cocoon in which the solar system has sheltered for 4.6 billion years is inhomogeneous, turbulent, and, to judge from the intrusion of supernovae, decidedly leaky.

References

1. Antonioli P et al (2004) SNEWS: the SuperNova Early Warning System. New Jour Phys 6:114
2. Church JA et al (2013) Sea level change in Stocker TF et al (eds) Climate change 2013: The physical science basis, Cambridge Univ Press, Cambridge
3. Dicke RH (1978) Is there a chronometer hidden deep in the Sun? Nature 276:676-280
4. Florinski V, Zank GP, Axford WI (2003) The solar system in a dense interstellar cloud: implications for cosmic-ray fluxes at Earth and ^{10}Be records. Geophys Res Lett 30:DOI 10.1029/2003GLO17566
5. Freeth T et al (2006) Decoding the ancient Greek astronomical calculator known as the Antikythera Mechanism. Nature 444:587-591
6. Kerr RA (2006) Who can read the Martian clock? Science 312: 1132-1133
7. Kollerstrom N (1992) The hollow world of Edmond Halley. Jnl Hist Astron 23: 185-192
8. Li KJ et al (2012) Why is the solar constant not a constant? Astrophys J 747, 135, doi:10.1088/0004-6378X/747/2/135
9. McComas DF et al (2008) Weaker solar wind from the polar coronal holes and the whole Sun. Geophys Res Lett 35:18,103-18,108
10. Mitrovica JX, Wahr J (2011) Ice age rotation. Annu Rev Earth Planet Sci 39:577-616
11. Mitrovica JX et al (2015) Reconciling past changes in Earth's rotation with 20th century global sea-level rise: resolving Munk's enigma. Sci Adv. DOI:10.1126/sciadv.1500679
12. Munk W (2002) Twentieth century sea level: an enigma. Proc Nat Acad Sci 99: 6550-6555
13. Nolan MC (1997) Poynting-Robinson drag. In Shirley JH, Fairbridge RW (eds) Encyclopedia of planetary sciences, Kluwer, Dordrecht
14. Reager JT et al (2016) A decade of sea level rise slowed by climate-driven hydrology. Science 351:699-703
15. Rettberg P, Rothschild LJ (2002) Ultraviolet radiation in planetary atmospheres and biological implications. In Horneck G, Baunstark-Khan (eds) Astrobiology. The quest for the conditions of life. Springer, Heidelberg, 233-245
16. Scherer K et al (2008) Cosmic ray flux at the Earth in a variable heliosphere. Adv Space Res 41:1171-1176

17. Souchay JJ, Dvorak R (eds) (2010) Dynamics of small solar system bodies and exoplanets. Springer, Berlin
18. Spada G, Bamber JL, Hurkmans RTWL (2013)The gravitationally consistent sea-level fingerprint of future terrestrial ice-loss. Geophys Res Lett doi:10.1029/2012GL053000
19. Steele JM, Stephenson FR, Morrison LV (1997) The accuracy of eclipse times measured by the Babylonians. J Hist Astr 28:337-345
20. Stephenson FR (2003) Historical eclipses and Earth's rotation. Astron Geophys 44:B21-B27
21. Stern SA (2003) The evolution of comets in the Oort cloud and Kuiper belt. Nature 424:639-642
22. Sturrock PA (2009) Combined analysis of solar neutrino and solar irradiance data: further evidence for variability of the solar neutrino flux and its implications concerning the solar core. Solar Phys 254: 227-239
23. Svensmark H, Eghoff MB, Pedersen JOP (2013) Response of cloud condensation (> 50 nm) to changes in ion-nucleation. Phys Lett A377:2343-2347
24. Taylor SR (2001) Solar system evolution (2nd ed) Cambridge Univ Press, Cambridge
25. Tobiska WK et al (2000) The SOLAR2000 empirical solar irradiance model and forecast tool. Jour Atms Solar Terr Phys 62:1233-1250
26. Vita-Finzi C (2010) The Dicke Cycle: a ~ 27-day solar oscillation. J Atmos Solar-Terr Phys 72: 139-142

Index

A

Abiogenesis, 80
Accretion disk, 34, 57
Alfvén, Hannes, 10, 30, 55
Alvarez, Luis W., 76
Alvarez, Walter, 76
Amino acid, 23, 58, 72, 82
Anaximander, 80
Antikythera mechanism, 85, 86
Archaean, 74, 75, 90
Arrhenius, Gustaf, 10, 30, 55
Asteroid
  belt, 20, 28, 34, 36, 52, 95
Astrobiology, 10, 80
Astrogeology, 9
Astronomical unit (AU), 4, 5
Atmosphere, 3, 27, 34, 49, 55
  Earth's, 33, 40, 44, 46, 56, 57, 74, 87, 88, 94
  of Titan, 24, 57, 73, 80
  of giant planets, 19, 56
  of Mars, 56, 57, 80, 90
  solar, 45, 91

B

Babylonian, 63, 85
Bang, Big, 16, 19, 24
Barringer, 76
Beryllium-10 (^{10}Be), 46, 47, 94
Bruno, Giordano, 1, 72

C

Calcium aluminium rich inclusion (CAI), 20, 22, 54
Callisto, 71
Carrington, Richard, 47
Cenozoic, 75
Ceres, 29, 52
Chamberlin, Thomas C., 8, 32
Chaos, 63, 66
Chaotian Eon, 74
Chemoautrophic, 81
Chemosynthesis, 74
Chondrite, 20, 22–24, 28, 43, 55, 57
Chondrule, 20, 22, 23, 27, 54
Clathrate, 18
Cloud, molecular, 6, 9, 13–15, 23, 24, 29, 54
Comet
  Churyumov-Gerasimenko, 57
  Hale-Bopp, 18
  Halley, 18, 19, 63
  Hartley, 20
  Philae, 93
  Shoemaker-Levy 9, 36
  Wild 2, 20, 22, 82
Comte, August, 8
Copernicus, Nicolaus, 1, 2, 4
Coral, 66
Core
  of molecular cloud, 13, 29, 30
  planetary, 32, 34, 42, 49, 51–53, 56, 58, 65, 87

© Springer International Publishing Switzerland 2016
C. Vita-Finzi, *A History of the Solar System*, DOI 10.1007/978-3-319-33850-7